Thirteen Lunations

Thirteen Lunations
A Celebration of Time

Khia Marin

13th Lunation Books

To the great and glorious web we call Life –
and each one of its members.

The Moon

The moon is like green cheese – just taste it, if you please.
It's fresh, and soft, and cheese.

As I went out one clear green night (was headed for the dairy),
I saw within a dimmish light that flickered soft and wary.
I thought it passing strange – very!

So I approached the dairy door, though feeling somewhat scary,
and saw within a manticore with eyes all wild and glary.
I thought it passing strange – very!

"Sirrah," says I "what do you here, all on this night contrary,
and fill yon pot with glowing sphere, and make it bubble merry?
You know it's passing strange – very!"

It said "You have such good green cheese, the best in all of Faery.
It asked me nicely would I please make it all bright and airy.
Is that so passing strange? Not very!"

The moon is like green cheese. Just taste it, if you please.
It's fresh, and soft, and cheese.

Contents

Foreword

Our Gregorian calendar feels dead to me; it always has. It has very little to do with what the moon is up to. The names of the months and days have become divorced from any meaning they may once have had. This book was born of my longing for a different calendar – one affording more meaning on a local level. The luni-solar one I re-discovered appeals to me. Owing to a change in how the weeks are organized, the very act of setting a date within it describes what the moon is up to and where the earth is in its journey around the sun.

A Nine-Day Week

I first met with the idea of a nine-day week fifteen years ago, in an online article that cited the early Celts. Their calendar was lunar-based. As they so aptly noted, the moon takes nine days to go from tiny to bright, stays brightish another nine days, and in nine more, gets practically invisible again.

The article next cited the Judaic calendar (still lunar-based), noting that they had a week like the Celts once, which changed when they were in captivity under Babylon. Babylon was in awe of the bright lights in the sky that looked like stars but moved strangely. Five could be easily seen. Their holy people gave each "wandering star" divine names and personae, assigning them each a unique day of its own to oversee in a repeating cycle. Then they added another day for the moon, and one more for the sun: seven in all.

The article (so seamlessly that I barely noticed) showed the order in which the Babylonian days flowed to be different from that which we use today (for our current days of the week are still

based on heavenly bodies), with no accompanying explanation. I
quite liked the older style, for it best reflects which lights appear
brightest as seen from Earth, and just as happens in actual early
morning and late evening skies, Mercury stays near Venus. Blissful
with this new window on older ways and long before there was any
dream of publication, I based my luni-solar calendar on that earlier
system, without quite being aware I had. Recently, I searched for
that article in hopes of citing it properly and found a very different
picture. It's in my Back Pages (pp 261-2).

We've learned a lot about our skies since Babylonian times.
We know the lights we admired are planets, and that Earth is one
of them. We even know of others – too distant to see, but still
near enough to be our neighbors. Why not make use of these
modern insights and by way of them complete the circle, returning
once more to that nine-day week so long ago forgotten? Building
on tradition, we could add a day for Earth, and another one for
those planets that are too far away to see, but are still counted as
members of our little group that dances attendance upon the Sun.
Our nine-day week, which should include three rest days rather
than two (I did the math), could go like this:

Gaiaday for the Earth: a planet and yet our home

Sunday for the Sun: sustainer of our life force

Moonday for the Moon and life's cyclical nature

Mercuray for Mercury: symbol of creativity and our
negotiations with fate

Vensday for Venus: attraction and inspiration

Tiwsday for Mars: struggle and right order

Jupitay for Jupiter, in olden days equated with a compass:
whatever is dearest, to help in navigating tough choices

Saturday for Saturn, whose famous rings symbolize the limits imposed by space and time

Starday for distant planets, unseen light and unacknowledged influences

But a lunar calendar would do more than simply add a few days to our seven-day week.

Lunar Calendars: Gregorian Translation

From Moonsighting.com I learn that lunar calendars start the first moment of a day when, at sunset, the new moon first becomes visible; for the new cycle begins the moment the new moon can be seen from the earth. On that day, you should be able to see the new moon for yourself, if viewing conditions are right.

Fifteen minutes after sunset go out and look to the west, close to the horizon. Location will vary a bit with time of year. Near an equinox, the moon will be just above the setting sun. During summer, it will slowly shift to the right and in fall or winter, go left again – always within 30 degrees of the sun.

When I use my calendar, if I wait for midnight to call the day new it doesn't feel as nice as if I do it at sunset. This personal choice helps me feel better aligned with the older holidays, yet it is private, with no impact on the flow of days as others perceive them. And when I choose the first day of the new cycle I wait for when the moon can be seen both locally and with the naked eye (some count the earliest time it can be seen, even if only in one part of the world or only through a telescope). I also want to know what time it is where I live when the moon emerges from the sun.

Stating time in local terms could make it seem as though the event is happening a day earlier near the Pacific Ocean than along the Greenwich Meridian, and waiting for days that best serve local visibility may feel as though the new moon emerges at different times on different parts of the planet, but these are just illusions:

tricks of changing time zones and varying calculation styles, like the ones discussed here.

So now you have a better idea of what a lunar calendar is. A luni-solar one stays in step with the seasons, too, by adding an extra lunar cycle (lunation) in certain years, as needed.

Dark Moon Time

During dark moon time, I avoid the use of names and dates, even though names for these days can be found (for example, First, Second, or Third Dark whatever the new lunation's name is). Instead, I keep things open. One lunation is ending and another beginning. The use of names does not feel right. I may be feeling this in sympathy with an idea we all used to know about contemplative practice (the concept came to me through studies long ago); during the short and variable time when the moon is hidden within the sun's glow we find a more powerful window for meditation or prayer. Now is the time to rest. Notice your dreams. Unfocused art such as collage, doodles, or stream-of-thought writing reveals unexpected depths. Enjoy these times. They are valuable.

Lunar Symbols and Time Zones

If anyone is wondering, a waxing moon looks like this:). It will keep getting thicker and brighter as it moves toward being full: O, and when it passes the full it will become a waning moon: (, getting steadily thinner until ready to disappear again.

UTC (or UT) means Universal Time: a time-scale based on the rotation of the earth as compared to celestial objects like stars and quasars. It uses a scaling factor to make cited times closer to Mean Solar Time at Greenwich, England; thus, it would hold true for anyone close to that meridian. PDT means Pacific Daylight Time. An event occurring at 20:00 UTC (8:00 p.m.) in France would be happening at 13:00 PDT (1:00 p.m.)(or, for very late or early in the year, 2:00 p.m. PST – Pacific Standard Time) in California, if it were being broadcast live.

Sample Dark Moon Page

Every box with a black circle in it denotes a day when the moon is so near the sun the one is hidden within the other's glow.

Times given here show the exact moment the moon re-emerges:

17:11:20 UT (UT: Universal Time):
(10:11:20 PDT; 9:11:20 PST)

This time the moon re-emerged at eleven minutes and twenty seconds past 5:00 p.m. at Greenwich; seven hours later than PDT and eight hours later than PST

A Calendar Key:

	Waxing Week
Gaiaday Luni-solar **1/17** **Su** Gregorian	
Sunday Luni-solar ○	18:38:40 7 Can 37
Moonday Luni-solar Rest day	If every third day were a rest day, we would work six out of nine days.

During **Waxing** Week, **Gaiaday** (first in the nine-day count) is on top, but during **Bright** Week, the position gives way to **Mercuray** (day four), and during **Waning** Week, **Jupitay** (day seven)

1/17: sample day falls on January 17th,
a Sunday   (Abbreviations for Gregorian days are:
Su, M, T, W, Th, F, Sa)

During **Waxing** Week, **Sunday** occupies this middle position; during **Bright** Week it would be **Vensday**, fifth in the count, and during **Waning** Week, **Saturday**

O, Symbol for the full moon (and for any day during which the full moon occurs), taking place here at 6:38 p.m. (and 40 seconds), in 7 degrees Cancer, 37 seconds

During **Waxing** Week, the bottom position is held by **Moonday**, during **Bright** week, **Tiwsday,** and during **Waning** week, **Starday**

Why all this Fuss?

The main purpose of this book is to demonstrate this resurrected calendar – how it would look as it moves through time. The left-hand pages give an approximation (if calibrated to a specific year they'd be clearer); the right-hand pages (seasonally apropos and serving up dreams recalled, adventures had, space science, folklore, poetry, history, and more) keep them company. If they were empty, there'd be journaling space for each day; filled, they took me places I never thought to go, dig at Europe's nature-friendly roots and in the end, tell the history of our western calendar. But why do I want this other one? I ask, I scribble, and still I do not know. Then I remember a trick I learned: *some* part of me knows what I'm up to, what all this lunar business is about. I decide to track her down and get her to talk to me (it's called an interview with the self).

I breathe deep and go still. Opening my imagination, I walk down the grassy path in my mind, across the open meadow where cool sunshine lies over the pink and yellow flower-speckled grass. The road slowly rises as it tops a knoll, then slips down over the other side. Ah, look there! Winging by above me: lovely birds, two – are they swallows? Maybe. I look back to the path as it dips down, lower and lower into a darkened vale. The light is a deep green in here and goes darker as I wind deeper in, dropping steeply between the trees. At last the path begins to flatten, curving to the left. It hugs a circular wall of stones, set deep into the earth and rising to a half-wall in height. This is a well; the water lies deep within it, and sitting upon the stones an older to middle-aged woman is resting, dressed in long robes of dark green. Pegged to the large trunk of an oak tree whose branches overhang the water and resting on the stones as well there is twine attached to a copper cup. She lifts it invitingly to me with a smile.

"Drink deep, daughter; may it cheer and revive you."
A ray of sun has poked through the trees as I lift the water and toss it up, splashing a bit all over myself. I am not thirsty, but it tickles me and I laugh, as the sun-ray turns the water to diamonds in the sky. She looks on, eyes twinkling to see me there at play.
"Mother, it's not for the water I have come to visit you this day."

"Do you think I didn't know? Yes, it *is* quiet in this forest... and the very silence speaks. I know you. But speak. Perhaps in the telling you can see more clearly what it is you want to know."

"Mother, I'm a mother too. There is a dream I bear, carried so long I wonder was it born when I was. Or maybe longer - it feels as though it were old as the hills. It is kin to the hills, the stars, the moonlight: a shadow dancing inside me as though among trees. Mother, why is it here? I know well, can see and understand well, that I want this calendar, want lunations and not months and want the names to honor the earth's changes as it journeys around the year. But *why,* and to what purpose?"

"Ah, child, to understand that you must be still, be still as I am still and give your attention to the place where you are. Go. Sit outside in your garden. Let the sun caress your shoulder, the moonlight fill your eyes. Watch the cats as they play with one another and laugh at the onion blossom moving around on its long stalk so, in search of the best place to scatter seeds. While you are there - *listen.* There is warmth there; presence too. You are no stranger to these presences, even if too absorbed to notice them. Always, you are welcome. It makes them happy to see you start to look around. One day, you might start *noticing.*

This calendar is like that - both invitation and invocation - to the larger world, maybe, but most of all, to you. There is so much beauty, so much love, such grace in the natural world around you and you seek to charm yourself out: out of your absorption, your busyness, your ever-running thoughts. They chatter endlessly, don't they? Like a running stream, only nothing like as peaceful. You *know* there is more, just beyond the shadow of your eyes! And so you come, you write, you sing this invocation. All this time, you thought it was for the world - and it is - but even more? It's for you. So go, take this celebration of time and look for others, those who are like you. There are many tales to tell. May they bless you, may they bless you all."

"Thank you, Mother. You're right. That is why. I'm going now. Slan agat!"
"Slan leat."

First Lunation

Riding the skies...

1

Storm Moon
Among Storm People

I had this meditation at a friend's house. Most of his life he kept an altar, and was very dedicated to the world of spirit. He had a mixed, mostly Native American background. When he meditated, he used his drum. His Navaho friend (a medicine man's son) may have been present at this particular drumming.

'The Kachinas came to our drumming. Having them in the room with us felt so good! One had a beak, and big glowing eyes, with what looked like flames radiating out around his head. One had evergreen branches in his hands and coyote fur on his back, with a tail that stuck out behind him. They came as paired dancers moving in a line. I wanted only to watch, but two gray and white striped clowns came and got me, so then we were three in the line. The one with the evergreens was in front, until someone with vines running through his hair came between us. They trailed down along the ground, and somewhere, there were squash blossoms. I had been sad, but being with them made me really happy.

We were going *somewhere*, spiraling out of our circle. Other imagery soon gave way to a chalky white path that climbed a steep, steep cliff into a golden cloud of light. Over the cliff and into the clouds a tunnel went through, but I drifted away. The vines helped, even so I kept falling, farther and further back. Though clinging tightly I still could not follow; it was like falling down hard, but without any pain. I let go and lay still, completely cut off, with strange black and white patterns filling my eyes. There I drifted until the drum became louder again.

They were returning, still dancing, moving triumphantly in a double line. They had brought the lightening, and they had the rain. I ran to them, shaking my own rattle like thunder. It felt like rain inside *me*, so clean and fresh.'

2

	Waxing Week
Gaiaday	
Sunday	
Moonday	In early Latin worship, the moon's dark aspect seems not to have had a name; perhaps they, too, were reluctant to use names in this mystical space. In later days when they needed one, they borrowed it from the Greeks (from Wikipedia). Dark moon time is the best time to cultivate silence. Some insights shine clearest where there are no words.

A Storm in the Mountains
This really happened – to me.

It's raining. Sometimes loudly on the roof, sometimes lightly blown across it, but never... truly... *stopping*. Khia couldn't remember a storm like this in all her life – but then, she hadn't lived out here as long as Bob had.

He was snuggled up damply between the covers, clothing hung up to dry, black hair curling slightly at the ends with wet. She looked up from her uninspiring desk. She wasn't freezing, only chilled through. She didn't want to admit it, would NOT burn up all the propane. At least the groceries have been put away. Peeking through the window at the private road ascending from the rural mountain highway leading home, all she can see is rain, falling like curtains across the sky. Bob's eyes are smiling over the top of the covers, and Khia's back is tired. He has the better idea. Nice dry covers. She abandons her desk. Nice cozy rain.

"I don't remember a storm this big ever. Do you?"

"No. Not since I was little, anyway."

"You think this'll bring us out of our drought?"

"Oh, no. We'll need a lot more than this. There was a little stream behind Chris's cabin that used to run all the time. It should run for a good few weeks now, though."

"Mmmmm..." Silence settles like an extra blanket for a time, each lost in private thoughts. Then, Khia giggles. "Remember when I was waiting for you to get off the computer, so I could surf the web? You left search results for my name on screen. That was such a surprise! I didn't think there was anyone else out there named Khia."

"Oh, sure there are. Plenty of people."

"Oh, yeah? Like, where?"

"Point somewhere." She takes a hand out from under the

	Waxing Week
Mercuray	
Vensday	
Tiwsday	On Mars in winter when one pole never sees the Sun, the surface grows so cold that 25–30 percent of the atmosphere gets deposited there. This and all other such astronomically-based blurbs are from information shared by NASA.

5

covers, points randomly at the ceiling.

"Hmmm, you're pointing to the north... the north end of Canada beyond the Arctic Circle, out in Inuit territory. Now, where's Khia? What's Khia doing? Oh. Yep, there she is." And Bob begins to tell a story he says is really a thought being mulled over by someone named Khia, living out in Inuit country.

The Blizzard

'Khia's lost. She opens her black eyes wide, but all she can see is snow – blinding, blurring snow. She's bundled up in a big furry parka, and *still* too cold, oh, so cold. She's been wandering around for what she's certain has been days. The sun never rose, but that doesn't mean much because it just doesn't, this time of year. Snow is all over her eyelashes, and on the curling tips of black hair that have snuck out in places from under the fur-lined hood. It's piled up everywhere, taller than her. She's never been lost like this in all her seven years. Her freezing little legs don't want to take another step, but she's smart enough to keep moving. She's really scared by now.

Suddenly she sees something, a few inches away from the tip of her nose, and learns what terror really is. It's a mother bear with some half-grown cubs! She's so cold she doesn't care anymore. She knows that one way or another, she will soon be feeling nothing at all. So she walks right up to them, holding her breath, and waits. The polar bear just sniffs her, pushes Khia into her warm white fur. Everybody snuggles down into one furry pile, and before they know it, they're asleep.

	Waxing Week
Jupitay	That Great Red Spot marring Jupiter's side is a storm with tumultuous winds peaking at 400 mph. More than three times Earth's size in width, it has raged for over 150 years.
Saturday	
Starday	

Storm Moon

The wind howls and snow goes arrowing along the ground – but not forever. As things quiet down, the skies begin to clear. When Khia awakes, the snow is gone from the tips of her hair. It has melted off of her eyelashes and she is all dry, rested and warm.

Filled with gratitude, she turns to the big white bear that took her in and says "Thank you, bear mother." A barely perceptible nod of the bear's head is all that dignity will allow, but Khia sees it. She wraps her arms around the smaller bear next to her and lays her head against his side. Not so grown up as they look, she and the young bears snuggle up one last time.

The world is white and strange, but she sees light reflecting off the snow, and feels a burst of joy. She can find her way home now. She can see, far away but standing tall, the village marker her people made out of earth and stone. Because of its shape, she knows just where she is – what side of town she'll be coming in on, where her house is, everything. Her little house by the big lake is the same as ever, but she feels like a different person: older, somehow stronger. She knows she is a very lucky girl.

Her village is amazed to see her. No one believes her when she talks about the bears. They just can't find it in their hearts. Everyone knows bears are hungry, all of the time. And they are wild animals. Why would a wild thing do anything but run, if by some miracle it wasn't hunting? No. It had to be the other way around: she was just lucky that she *didn't* meet those bears. If she had, the village would surely never have seen her again.

It hurt Khia's feelings that no one would believe her. It wasn't fair, being disbelieved because of your age. They knew *she* believed what she told them, but they thought it was a dream, brought on by exhaustion.

	Bright Week
Gaiaday	
Sunday	99.86 percent of the mass in our entire Solar System rests within the Sun, and it is the gravity of this massive Sun that holds the rest of our Solar System together. A million Earths could fit inside the Sun.
Moonday	

Silvery days chase each other through the cold, and begin to lengthen. A warmth, barely felt, begins its slow dawn across the land as bits of green appear. Khia is eight already. Now the days are long indeed – warm, sunny green worlds, where it's hard to get any sleep.

One day, some visitors come to town – large, furry white ones, snuffling the breeze and strolling down the middle of the street. Terrified, the villagers scatter, but Khia recognizes her friends. They might have grown a bit – okay, a *lot* – but they're the same goofy cubs that shared their nest so trustingly with her last winter. She doesn't see their mom. Has she left them on their own now? Khia can't hold back; she runs to greet them. One good hug, and she's scratching their shoulders, tickling their bellies, rolling with them on the ground, wrestling and playing just like she did when the snow stopped and everyone was so happy to see again. She and the bears are that happy now – happy like that, seeing each other.

The villagers just stare. They're terrified, and her mother is furious (for a little while). Then they are simply stunned. One by one they start thinking about what she told them when she came home last winter. Could it have been true after all? How do she and the bears still know one another? There they are, playing like friends together right before our eyes. What does it mean? It's strange!

Khia's having too much fun to notice anything else, but after a while, the bears move on. Khia looks around for her human friends again, and sees everyone in a cluster some distance away, watching the bears leave with eyes like saucers.

"H-how did you d-do that? C-can you t-talk to bears?"

"We made friends. Don't you remember how they kept me warm? It was so nice to see them again!"

	Bright Week
Mercuray	"Astrologically speaking, Mercury is your lens, the patterns of your thought, and the shape you give reality." It can even be intellect itself, loving to take things apart and put them back together again. Wanting to give credit to whoever originated the quoted portion of the above phrase, I did a search for it and got nearly 7 million results!
Vensday	
Tiwsday	

"So what you told us must be true after all. But it's so strange! Isn't it strange?"

"No, I think *you* guys are strange. Did you really think you knew all about bears? You were so sure about it I had to be lying, even though I had no reason to lie. Don't you think *that's* strange? I sure do!" '

Bob told me several stories that season, because the rains went on and on. He said they were just thoughts that had captured the various Khias' imaginations. One of them lived in China. She was dreaming about a day when she and her fiancé had gone to the mountains in honor of a certain custom. Lovers would climb to find a large collection of locks, add theirs on, and lock their love in. Bob found this in Chinese Khia's head one rainy day, though he had never heard of it before. I searched on the web, and learned that the tradition does exist. I don't know what to think about it all, but I do know I was grateful for the entertainment afforded me, in those days when there was nothing to do but try to stay warm and wait for the world to dry out.

Clouds

Wild white shining wings,
legs outracing the wind,
breeze-wispy, high-arched tails:
I have seen them!
The children of Pegasus.

	Bright Week
Jupitay	
Saturday	Balbus asserts the name Saturn comes from Latin satis: "sated" (with years, and more - in the Roman mythos, Saturn represents Father Time, who, at the last, consumes all things). This and all such blurbs based in the Roman mythos are from linked Wikipedia articles found while investigating planetary names.
Starday	13

Storm Moon
Epiphany

On January 6th (Epiphany, or Twelfth Night), the Christmas season ends and Carnival begins. This is a western Christian festive season, with many kinds of public celebrations, such as parades, street parties, and elaborate costumes with masks. It's a time for mock battles, food fights, social satire... for reversals of traditional norms, when language and behavior normally forbidden is indulged. It happens more often in localities with a large Catholic presence (like Mexico, Brazil, Italy, or Spain).

January 6th marks the day the three Magi, following a mysterious star, came at last to baby Jesus. They were led 12 nights through the desert, finding him when he was just 12 days old. "Epiphany" comes from the Greek and means "to show."

In New Orleans on this day, a Society Ball is held, and costumed bands of Twelfth Night revelers hold their annual rides on two different streetcar lines, singing and joking all the way. Joan of Arc's birthday is celebrated, with historical characters in medieval dress parading through the French Quarter. It's all about mystery, magic, and spontaneity. New and unexpected pranking of every kind is heartily appreciated.

Spain and Mexico celebrate this day as Three Kings Day, when the three kings brought gifts of gold (for royalty), frankincense (for a divine life), and myrrh (for death's defeat). On this day, Latin children receive gifts, too. They leave their shoes by the door so the three kings can fill them up. As a welcome, some families leave grass or salt for the camels.

Parades and performances often take place on this day. In Mexico City, bakers make a sweet pastry a mile long that represents a king's crown. People fill the streets to get a slice. It's considered lucky to find a baby Jesus doll hidden inside.

	Waning Week
Gaiaday	The word "earth" comes from an Old English/Germanic word meaning "the ground."
Sunday	
Moonday	

15

Storm Moon
A Hard Fall

I took a hard fall one day and bruised my shin-bone. My doctor said I was not to put weight on it until it had healed, but he couldn't tell me when that might be. I asked around; I never could learn when it would be safe to expect normal functioning or how I would know, so I waited carefully for it to normalize on its own. It got really hard to look after myself. Everything I needed was far away, and up or down the side of a hill. This went on for far too long. One day I couldn't stand it anymore and ran away to my mother's house. Conveniently, she and her husband had come to a town near me for medical work. I went home with them, expecting to be there only for a few days.

It started snowing soon after we arrived (they lived in the hills too, but far from me). He started having strange pains soon after what should have been minor surgery. In the end the two of them drove to the hospital in a town near them, but they almost decided to wait a day and see if it would clear up on its own. The snow would have been so deep by then they couldn't have gotten out, not for a week, at least. As it was, they weren't able to get back *in!* But this didn't matter, because the snow had all melted before the infection he was fighting had been safely subdued. It was a dangerous infection and there was nowhere my mother would rather have been than by his side.

Safely at home, my grandmother and I looked after each other. We visited, watched movies, and listened to music. Then I got an idea. I'd always wanted to make snow ice cream, but we never get snow where I live. And here during my visit were acres of it, right outside the door! So I did some exploring and learned all I needed to know. The trickiest part about making ice cream from snow has nothing to do with what you can buy at the grocery store. It's about the snow itself. Where you get it from and what you do with it will have the biggest impact on the quality of ice cream you can later create.

	Waning Week
Mercuray	
Vensday	Venus has a very weak magnetic field. This was a surprise to scientists, for unlike Venus, Earth's is powerful, sheltering life here by deflecting many of the charged ions spinning off from the Sun, and is generated by the nickel and iron at Earth's core as it spins rapidly in space. Perhaps Venus's field is weak because its core is molten and not solid like Earth's. It never, ever cools!
Tiwsday	

Snow Ice Cream
(best with fresh-fallen snow before the sun comes out)

Choose a container for the ice cream and pre-chill it in the
freezer at least six hours ahead. Before you take it out, find a
deep, shady snow bank. Now go get your container and embed it
in the snow. Brush off the top layer and fill the container with
clean stuff from the center. Put it back in the freezer, while you
get your other ingredients ready. What flavor do you want? We
used chocolate syrup, and I could have added crushed candy
canes. Mashed fruit is delicious. Your imagination is your only
limit.

Here's the basic recipe:

> 1 can condensed milk
> 5 cups fresh-fallen snow
> 1 tsp or more of any flavorings

Alternatively, you could:

pour 1 cup milk, ½ cup sugar, 1 egg, and your chosen flavors into
a bowl, beat them frothy, and add enough snow to absorb the
liquid.

I loved my long visit with my grandmother. And one of my
mother's friends was an EMT. She was able to tell me my leg
had healed well and was safe to use again. I'd traded crutches for
a wheelchair the day I arrived; now I joyfully used it less and less,
as my leg got stronger and stronger. What a blessing for all of us
that visit proved to be!

	Waning Week
Jupitay	
Saturday	
Starday	The word "planet" comes from Greek "planetai" (asteres planetai: wandering stars).

Storm Moon
The Word 'Thursday'

Thursday takes its name not from Norse Thor but from Anglo-Saxon Thunor, though both are associated with weather. Blacksmiths felt a special connection with this god. They believed the sound of thunder was himself striking his hammer on his mighty anvil and lightning was the spark created by the strike. The hammer (short-handled, and useful as both a held and a throwing weapon) was his symbol. Pendants showing such have been found in many Anglo-Saxon graves – often worn in praise, much as we wear crosses in honor of Christ today. He was a god of simple physical strength, no doubt warmly held by simple folk, who worshipped in groves and in meadows. The Romans equated him with Jupiter, their own thunder god. While doing this search, I learned that certain heathen kindreds exist (at https://greatvalleykindred.com, for example). It was their works that helped me get the best understanding of the Anglo-Saxon gods for whom our days of the week are named.

In Astrology: The Two Wheels

The Wheel of the Houses is drawn from the way the hours pass on account of the Earth's rotations around its own axis. The Zodiac is a different wheel, taken from the way constellations forming a background behind the planets change throughout the year as the Earth circles the Sun. The two wheels make it easy to describe where a planet is in time. For example, if we say that Venus is in Taurus, we mean the constellation of Taurus was in the background at the time in question, and if at that moment Venus was in the Tenth House, that's how far or near it was from zenith or horizon.

Second Lunation

		Between formless and form...
●		
●)		
		21

Dark Moon
A New Year's Celebration

This amazing series of meditations came several years ago in early February, so strong and clear it felt as though I were really there, in the company of real people.

2/7/14 *Gardenia*

I see a luminous gardenia floating on water and being reflected in it. Is it a gardenia or is it a candle? It has a lovely smell, of – what? Ah, I remember: a walk through a forest, spicy... indescribable! I am ready to pluck it so I can stare at it more closely, but then I try going inside and find all my friends there, radiant and small. They lay me down in the nectar flow so I can drink too, and I drift off into the most peaceful stillness...

2/13/14 *New Year's Eve*

It looked like a white flower, all lit up from within. It was in my hand; the warm glow of the light was on my face. Then it went off with all the others, afloat, reflected on the waters. Spicy smell – now why is that familiar? Ah, the grove of trees... Side by side, Silvertree on my right and Rumiel on my left.
  "These are our wishes."
  "These are our blessings."
We release the old, make our prayers for them as we let them go – troubles we've made peace with, struggles not relevant to us anymore. We are ranged along the shore of the lake, all together, faces illumined by the small flame burning before each of us. The boats are of wax, very thinly carved and very white, floating on the water. We each have several.

	Waxing Week
Gaiaday	The word "Gaia" is actually the name of a lesser known Roman goddess, seen as the ancestral mother of the interwoven web that is all life.
Sunday	
Moonday	

"Now the old has burned away. Now the new can come and play. With our wills we let love stay, all the heavy float away." Our hands are above our heads, like flames – we *are* the flames. We burn and release and bless all those things that are ready to be let go. And now it's back to our blankets, under the trees, to sit in small circles with the people we love and tell each other all about it. There are hot drinks – cocoa, cider, spiced wine – and food, too, lots of things... buttered potato bread, roasted meats and veggies. Good to replenish, after all that play. Happy New Year, everyone. May your year be blessed, and filled with love.

2/14/14 New Year's Eve Night

First, velvet darkness, then, the forest at night. Dancing fireflies; people with me and the happiness of feeling connected. There is a large, hollow tree; we are all filing into it. Big circle, soft piles of hay – we're gonna party all night long. That's a New Year's tradition too: all the heavy things got burned, and now everyone comes together to celebrate what's left. Not in so many words – just being together, telling stories or jokes or drifting off to sleep while someone else shares and other people listen or laugh. Singing if you want to – everyone's there, even babies and their mamas, being together and sharing in the dark. Oh, and a timber wolf dropped by for a visit. This is a time for friends, fun and spontaneity. Short time outside – hugging the big tree gratefully, enjoying the luminous peace and silent beauty. Outside, but not left out. Not ever. The moon is full, as in my waking world; it's the first full moon after the Chinese New Year.

2/15/14 New Year's Morning

Walking among dangling vines hanging down under the trees: white, four-petalled flowers, golden pollen – letting the flowery vines trail

	Waxing Week
Mercuray	It was Greece that gave us Mercury as a messenger. He is still associated with messages (post, email, and phone, even dreams or divination) and any form of communication, such as writing or speech.
Vensday	
Tiwsday	

25

over us as we walk. We've spent all night awake, inside that tree, laughing and telling stories, dozing and dreaming, snuggling and wishing: "Wouldn't it be lovely if?" and "How about this?" Now it's broad day and we've come out of the tree, are walking under these dangling vines. Our heads bump the flowers and send the pollen out in explosions. We walk mindfully; we are listening, have been listening all this time. Pollen explodes. Ideas move. We come to the edge of the trees at the top of a gentle hill, where mid-morning sun shines brightly and we hurtle ourselves over, prone, rolling down and down, shouting out our ideas all the way: "Pipelines carrying water away from flooded areas and off to drought-stricken ones!" "Community gardens in all the city parks!" (My ideas – I couldn't hear theirs too well while rolling down the hill!) We all land in a jumble at the bottom, in a warm pool of sun on a chilly morning. Most of us just lie there, and doze. Perhaps there will be dreams for some of us. Whatever is beautiful, whatever is inspiring in any of those dreams will be made into pictures and used throughout the year. They will help us remember these times, when we listened for the love the universe wants expressed and committed our bodies to helping that love be born.

2/18/14 Make a Wish (New Year's Night)

Sitting with a candle in my hands: white, fragrant, another hollowed out flower candle. I'm staring at the flame, seeing it reflect in the melted wax, watching sparks fly in the air. I sit with the others, around a campfire. Did we party yet another night? Knowing us, we probably did! Dawn is not far off. This candle is a personal one. It's for making a wish, or telling someone's fortune... or letting someone tell yours!

	Waxing Week
Jupitay	
Saturday	
Starday	Metaphor and science are separate now. We know planets are not stars, and even stars – while amazing as powerhouses of heat and light – are no more divine than your great-aunt Mabel (yes, to current ways of perceiving, Mabel is after all *not* divine).

New Life Moon

Kindly Ways

Water bugs, sunbeams, wander down the river's gleams,
wander down the river's dreams
of sun and star and kindly ways.
River trance, heron's glance – sparkle, lively radiance!
Stay to see the heron dance,
and dream of stars and kindly ways.
Ancient singing, ya'at'eeh: Night Chant, bring forth the day!
Catch the web of flow and sway,
and bend and turn in kindly ways.

Spring Feelings

Only yesterday it was rainy, cloudy, dark and cold, but now the sun is out a lot: new life is stirring, and we feel a stirring too. I feel inside the coming of Spring (wide-eyed, golden-haired girl, encircled by children), even though they say she isn't here.

Candlemas Day has come, is coming. I read about this day, seek a fitting introduction for those amazing meditative experiences I had, and find instead stories that waken echoes in my soul. Behind this picture of Spring, nearly erased but echoing within me still are symbols I didn't realize were connected, but they are: symbols carrying a mystery I can feel. I learn from Mara Freeman (out at Chalice Center) that it is Brigid, ancient, pre-historic Briga of the Brigantii and Pretanic tribes, around whom these images resonating within me cluster.

From the Pretanic people comes the Welsh word for Wales (and at one time, all Britain): Prydain.

	Bright Week
Gaiaday	
Sunday	
Moonday	Luna was but one aspect of a triple goddess in Roman Italy, divine reflection of the moon and its three phases: crescent, full, and dark (Maiden, Mother, and Crone).

And from this same root the word Britain comes too. At one time the Brigantii were all over Europe. They left traces of their name in Breconshire (Wales) and also in Brechin (Scotland). Bregenz (Austria), was their capital city. When the sun began to set on their holdings and they withdrew into a more closely collected place, it was in Northern England that they mostly held sway. They had a Heavenly Mother instead of a Heavenly Father, and they called her Brigantia (Briga, Brigha, Brighid, Brigid, Brigit). Somewhere deep inside, it seems that a part of me has not forgotten.

The following poem, inspired by a silly rhyme, I kept because in me it captured a sense of wonder. I didn't know why just then. And now? I suspect I do.

That Cow of a Moon

Hey diddle diddle, who's that on the fiddle?
Where's that cow of a moon?
Through the fields she rides, her *lyre at her side,
playing the oddest, saddest tunes.

The dreams she weaves for sleepers like these
won't be forgotten soon.
No, they'll burn in the brain like hope come again
in the strange, sweet songs she croons.

Where she might go no one can know,
for she's blown, like a heaven-sent boon,
on the back of a steed adrift, without heed
for aught but itself, **Aroon.

	Bright Week
Mercuray	As seen from earth, Mercury and the sun are so close together that Mercury is only visible to the naked eye for a brief time: after sunset, before it follows the sun beyond the horizon.
Vensday	
Tiwsday	

31

In its wild, dark heart lies the bubbling start
of luminous truths, all known
without knowing why, but they bring us a sigh,
and a smile, as our dreams we hone.

Oh! 'Twixt curving horns, the softest light's borne.
It brightens the crests of the dunes,
as they amble along, trailing echoes of song,
fresh from the weave-dreamer's loom.

*A lyre is a stringed musical instrument in use long ago. **Aroon is Irish for darling. To croon is to sing, a boon is a gift, aught means "anything," to hone is to sharpen, 'twixt: old word for between, as "borne" is for carried.

St. Brigid's Day

Of Brigantia, little is known, but of St. Brigit, much. No proof can be found that such a person was ever born, nevertheless, stories of her person are told in plenty. Perhaps the people invented her, so they could talk of their beloved goddess under their conquerors' noses?

"In Ireland" (says Mara Freeman) "Brigid was believed to travel about on her special day, blessing people and their livestock, and so an offering of cake or bread and butter was left outside on the window-sill for her. Sometimes they left a sheaf of corn too, as sustenance for the white cow that traveled with her. Or a bundle of straw or fresh rushes were laid on the threshold for her to kneel upon to bless the house, or possibly so she – or the cow! – could wipe their feet before entering. The cow supplied her with all the milk she needed – every day lakes of it.

	Bright Week
Jupitay	
Saturday	Taking a walk in Roman mythology, Saturn poses the question: what will you do with your time on earth? It is associated with precision focus, dedication, productivity, great achievements, lofty goals, and career.
Starday	

She fed birds, animals, and the poor; turned water into ale and stone into salt."

Both wells and fires were sacred to her. She had a fire in Kildare that burned for a thousand years (from the fifth to the sixteenth centuries), in an abbey surrounded by a withy hedge which, Gerald reports, "no male may cross." She cares especially for hearth-fires: "the power of the sun brought down to human level" (Mara Freeman: www.chalicecentre.net)
In the Hebrides, Bride (Irish Brigid) is called Mary of the Gael. The people like to think of her as the midwife at Christ's birth.

Candlemas

Today the Day of Brigid (in Ireland, or Bride in Scotland) is equated with Candlemas and the Feast of the Purification of the Blessed Virgin Mary. In the church's understanding of Jewish law, Mary had to be purified after giving birth to Christ, and was only allowed to come out of seclusion after forty days had passed. As a young girl, it made no sense to me that giving birth to God's own Son could cause a person to be called unclean. It rankled and I used to resent it some, but I see now that it all depends on the spin one puts on these laws. The same laws with a different spin will have a different meaning entirely. What a blessing it must have been, after the long hours in labor and with the precious, fragile new life in your arms, to be spared the hard work it was necessary for everyone to share in at that time, while you heal, bond with your baby, and master the intricacies of nursing and caring for him! This was a kindness that strengthened the entire tribe.

	Waning Week
Gaiaday	Even on land that was once desert, wherever the trees grow, farming can resume. Keeping chickens and other animals helps rebuild the soil too. Doing these things has led to the re-greening of as much as three million hectares of land in Niger. I learned this by asking YouTube whether desert land can ever become green again.
Sunday	
Moonday	

New Life Moon

Now is the time when ewes give birth to lambs and their milk flows. It has been called Groundhog Day, Lady Day, Candlemas Day, and in early Celtic thought Imbolc (meaning "In The Belly": a day closely associated with milk, water and birth).

Spring Tonics

This time of year has many associations with purification as well. The holiday feasting is over and New Year's resolutions are still strong. Fresh greens have only just grown in again, with plenty there now to do a body good, so what better time for a detoxifying cleanse? And if it should happen to taste nice?

At www.Solasbhride.ie in Kildare (where St. Brigid is still celebrated) I learn that "**Nettles** are a traditional spring tonic – an ancient Irish way of flushing toxins from the system that actually has some scientific basis. Nettles are rich in iron and also valuable in the treatment of arthritis. Pick only the tender tops (never use tops which have flowered) and use gloves when picking them. It is possible, by harvesting frequently, to keep young nettle tops going right through until the early summer."

They shared there a recipe that tasted delicious when I cooked it with **watercress** (which is lovely for a person's liver, and was a little easier to get at the time). I have it right here, but it will look all the better for being presented in one whole piece on the following page. While we're waiting, let me just say that **cactus pads** too have been relied on to cleanse the system, and they are especially helpful to people who eat a diet with a lot of grease in it. Just pop them into the blender with some water. My neighbor fed his reluctant nephew some cactus pad tacos, and the nephew liked them quite well.

	Waning Week
Mercuray	
Vensday	
Tiwsday	March was named for Mars. Festivals were held in March when farming would begin, and again in October, when farming ended and military campaigns would start.

New Life Moon

Aha! Here we are:

Spring Tonic Nettle Soup

Ingredients:

2 cups potatoes, peeled and cubed
1 cup mild onion, peeled and finely chopped
3 cups closely packed nettle tops, washed and roughly
 chopped
2 Tb butter (or bacon, duck, or goose fat)
6 cups chicken or turkey stock

Method:

1. Melt the butter in a large pot and sweat the onion and potato for about 10 minutes over a gentle heat.

2. Add the stock and bring to the boil.

3. Wash the nettle tops, drain them and add them to the pot to simmer for 5 minutes only (don't let the bright green color fade or a rather strong taste will develop).

4. Test for tenderness – don't worry, they don't sting after cooking!

5. Purée the soup until smooth in a food mill or food processor. Return to the pan and reheat.

6. Serve garnished with a swirl of cream and 2 tablespoons of either parsley or chives, chopped fresh.

	Waning Week
Jupitay	
Saturday	
Starday	Astronomer-astrologers of old believed the wandering of certain "stars" against the background of stars that stayed put portrayed in living light the direct movement of gods among human affairs.

39

New Life Moon
Planets: Rejoicing, in Fall, and in Exaltation

I learn at https://demetra-george.com that our world once used a
system now forgotten, where each planet was the joy of a specific
house. A **rejoicing** planet gets strength from its happiness and can
overcome almost any challenge it may be facing at the time,
moving events towards joyful outcomes that match its own
condition. There were only seven known heavenly bodies then,
each with a house (or part of the sky) to rejoice in, with names
like Life, God, the Goddess, the Good Daimon, the Evil
Daimon, and Good Fortune, or Bad Fortune. When a planet
was in the area it loved, it was very happy and much more likely
to be its best self because the beloved environment surrounding it
helped it stay positive.

Concerning rulerships and falls:

- When a planet is in the constellation it "**rules**," that means it
  is home: from within these surroundings its actions are most
  likely to play out in the best way.
- A planet "**in exaltation**" is in a sign that's almost like home
  (like visiting a relative or close friend): not quite the same,
  but help and happiness are there – the influence is positive.
- When "**in detriment**" a planet is in a sign that's the opposite
  of home (they're always arguing and getting in each other's
  way): action may be misused, negative, poorly formed or
  blocked.
- A planet "**in fall**" is in a sign opposite its exaltation, so it
  won't express itself as well (there will be less strength).

A **peregrine** planet is not in any of these signs. In two cases out of
three, planets *are* peregrine. To modern astrologers, aspects
(angles planets form with one another) matter more than
rulerships or falls.

Third Lunation

	Between born and unborn...
●	
●)	

41

Dark Moon
A Healing Experience

I had a wonderfully healing experience during a meditation one day. I was able to feel, in a direct and almost physical way, the action of the power of the blood of Christ. I wasn't feeling very good that day. I felt wrong, off, dirty somehow. I was being defensive in a hundred little ways that made no sense to me. I didn't know what was wrong, but something was out of sync and I just didn't feel right. So I went off somewhere quiet and went into meditation. It was dark and quiet and very comforting, and then images began forming in my mind. I was walking along a path. In a short time, I approached the edge of a pool. It was bordered in stone, with stone steps leading down and in. As I was descending the steps I felt the water flowing over my feet, higher and higher as I went down and in. It had a lovely penetrating heat that went right through me. In color it was a beautiful violet-crimson that was always changing – a bit like fire. Somehow the idea came to me that in this pool I had found the blood of Christ, which was also a flame; cleansing, warming and healing but never doing harm. I felt a deep gratitude, and when I emerged, I felt whole and clean again – somehow new.

A Discussion

I had the privilege of being part of an online dreaming group for a time. We paid attention to the dreams we had, watched for any that came at special times, and shared whatever we felt to be relevant. I was grateful to be assured no one would mind if I share here parts of a conversation shared there, as I feel it would greatly enhance these pages.

"One of my faculty colleagues sent the following message out about this last week:

	Waxing Week
Gaiaday	This is Gaiaday of Waxing week: Waxing Gaiaday, the first in each lunar cycle, best for trying some thing new or doing an old thing in a new way. There is a meditation for such days that feels nice when done solo and even nicer as part of a group: sit in peace with a pale blue or silver candle and start thinking "wouldn't it be lovely if...?" Embrace your dreams; name them all.
Sunday	
Moonday	

Spring Moon

'According to the Babylonian Talmud (B'rachot 59b), every 10,227 days (28 X 365.25 days) the sun returns to the position it occupied when the universe was first created. The tradition holds that the sun was created in its spring equinox position, at the first hour of the night before the fourth day of Creation. The next opportunity to recite Birkat ha-Chamah will take place on April 8, 2009 (14 Nisan, 5769). We'll gather – rain or shine – for a brief ritual in the parking lot of Menauhant Beach (East Falmouth) and to recite the Blessing for the Sun. We'll bring coffee and you can bring any tasty chometz (leavened bread) you're trying to get rid of (as Passover begins that night).'

My response to this message follows:

'This is interesting! It does not accord with any solar positioning I am aware of - but 28 years is equivalent to a Saturn cycle. There is a good deal of evidence for associating the old Canaanite god El (who later becomes Eloah and Elohim) with Saturn, and his son the storm god Ba'al with Jupiter. There is also some evidence to the effect that the new year used to be celebrated at the time of the Spring Equinox in Judah and Israel, and that the move to celebrating it in the Fall is of Talmudic age only - perhaps to separate it from the Babylonian (pagan) New Year's festival which also traditionally fell on the Spring Equinox.'"

"Fascinating! Since this spring full moon is also especially beloved of fairies, I see a wonderful convergence of traditions, here! Thursday dreaming means the dreams which one remembers on waking Thursday morning, so the dreaming of course begins tonight, in correspondence to the events you describe below.
This night is also Tenebrae for observant Yaquis, which involves going into and facing the darkness to clear the way for

	Waxing Week
Mercuray	Mercury was known in Rome as "keeper of boundaries." He had a role as bridge between the upper and lower worlds, guiding souls as they left one for the other.
Vensday	
Tiwsday	

45

the light of Easter. So we have two ends of the spectrum here – one community blesses the sun and the other clears the darkness out of the way of the light. Who knows, then, where our individual dreaming might take us? Blessing the light? Battling the Unseelie Court? We shall see! Whatever it might be, it sounds like exciting times."

"Can you describe the flowers? I have heard of the tradition about the relationship between the blood of Jesus and the flower, but specifically with reference to the dogwood, which blooms around Easter (and which is blooming magnificently in my area right now). The dogwood has a white flower with 4 petals, resembling an equal-armed cross, with a red center."

"Flowers have intensely charged meaning for Yaquis. They bridge the worlds of Christianity and the Sea Ania (a Yaqui world of Spirit and of Nature), in a number of ways. In the Christian context, Yaquis believe that every drop of Jesus's blood that hit the ground became a flower, and that flowers are, in church, sacred symbols of His baptism of earth and everything in it, and also of love, prayers, and grace.

In the final enactment of the battle between good and evil before Easter, the Gloria, men in diabolical masks charge the open-air ceremonial chapel, only to be conquered by good guys throwing flowers at them– throwing at them love, grace, and prayers, and the healing power of Jesus's sacrifice – which causes them to destroy their masks of evil and regain their real faces.'"

Flowers mean so many things, have so many beautiful associations. Twice in my life, without realizing what day it was, I've written poems (or they wrote themselves) in honor of spring. Both were written on the very day of equinox. One made the memory of a certain flower the leading subject of its ruminations; the other was all about the day itself. It carried the feeling of

	Waxing Week
Jupitay	
Saturday	
	Saturn's enclosing and surrounding rings reflect the idea of human limits such as those imposed by a person's sense of duty, ethics, civility, commitments, or responsibilities, as well as tradition, hierarchy, authority figures, practicality, reality, and time.
Starday	47

traditions long forgotten, experienced as freshly as if they happened yesterday. I should like to share it here. The first verse is for daylight and the second (not arriving until many years later) is for night.

Spring Tides

Circling sun:
spring is come –
flowers on the hill!
Songs, and laughter...
Follow after,
pick your fill!
Pick your fill.
Sparrows fly,
larks cry:
the tide of life
is high, is high.
Beat the drum!
And come, oh come –
chase me up the hill!

Circling sun:
spring is come –
fires on the hill...
Songs and laughter follow after.
Take your fill! Take your fill.
Sparks fly,
doves cry:
the tides of life
are high, are high.
Beat the drum!
And come, oh come,
chase me down the hill!

	Bright Week
Gaiaday	
Sunday	
Moonday	Diana (from dyew: 'sky', root of Latin deus 'god' and diēs 'day', but as an adjective, so "bright like day") had strong associations with the moon.

Spring Moon
Bunnies and Eggs

The moon has been called the "open-eyed watcher of the skies." (see Notes, p. 263) Hares too are born with their eyes open. Long ago, they were associated with the moon. As a child, even I had heard of the Man in the Moon, but when I searched the full moon for him, all I could find was a rabbit. It had very large ears; perhaps the "rabbit" was a hare.

You've heard that phrase: "Mad as a March hare"? In March, it's true that they behave strangely. They twitch and jump at the least little sounds. This is no mystery: springtime is breeding time and they've *all* gone mad (for love).

Of old, there was a strong connection between birth, sex, and spring. Rabbits (and hares, especially) embody these ideas so well. Could these mad March hares have melted down into our own Easter Bunny? As children grow, they learn that birds and even snakes produce eggs, but never rabbits. It's a self-negating idea, safe enough even for non-Pagans.

Nevertheless, a second look might prove rewarding. Long ago, eggs too were filled with meaning, as symbols of new life (the place from which new life emerges, and that which sustains whatever is new and fragile until it becomes sufficiently established to be ready for this physical world).

In medieval times a feast at Easter would have many flowers on the table; on one side a huge golden sun would be placed and on the other, an enormous silver moon (in full, not crescent phase). The mood would be both solemn and joyful, irreverent and pious. Here too, there are eggs of which much are made. Then as now they are boiled and decorated: painted beautifully, with their ends bordered in lace, or tiny glass jewels. They fill a bowl to the brim in the center of the High Table. High-ranking families would use designs and colors unique to

	Bright Week
Mercuray	
Vensday	
Tiwsday	Mars was said to have been born on New Year's Day (March first in early Rome: "the Kalends of March").

their houses, and give the eggs (Paschal or Pace Eggs; pasch is a Jewish word meaning both Easter and Passover), as gifts to entertainers who pleased (see Notes, p 263).

Decades ago, I decorated and gathered Easter eggs with a family living in the Navaho Nation. Maybe it was because the family was so large? For whatever reason, it was a truly magical time. They bought and boiled many dozens of eggs. Everyone was welcome to help, and besides an adult or two there were a lot of kids who got involved. The kids were amazing. They kept their thoughts relaxed and open. Ideas flowed freely in a delightfully unfocused way – we dyed and decorated the eggs not just with pictures, but with ideas, hopes, dreams, fears (well okay, words), too. Suddenly, those eggs weren't just pretty; all sorts of new energy waited inside.

Before the hunt, everyone stayed inside the house except for one sister's husband, who took the eggs outside and drove around, looking for the right place on the grandparents' ranch to hide them all. When everything was ready, he came back for us, and showed us where to stand. We were all lined up in a row, and on the count of ten, we started gathering. There were so many – easy ones for little kids, tougher ones for folks who like a challenge: in the end, several dozen for each brother's or sister's family. Those eggs were delicious, and they were also potent – with new ideas and doorways into new dreams.

Among the Teutonic tribes the end of winter never came until around the spring equinox. They celebrated its ending by kindling new fires on Easter Day. You still find this idea of newness in a lot of families. In spring we, too, may feel as though the whole world is new. Indeed, we can feel this way any time of year, with help from the love found in Spirit and the power of Christ's gift to us. It may be just inside us, but still, it changes everything. On Easter Day we celebrate

	Bright Week
Jupitay	
Saturday	
Starday	Uranus is associated with unexpected changes, unconventional or radical ideas, ingenuity, originality, discoveries, democracy, and revolutions. It governs any gathering based on humanitarian ideals or revolutionary events that upset established structures.

that sense of the new and see it's reflection in every lovely spring day. On this day especially, wear something new.

On Easter Morning

It is Easter morning,
*   and Mary is walking*
*   to Jesus' grave,*
to anoint him
*   with tears and oil.*
The sun had risen.

Now
*   She comes to the grave*
*   and sees the stone:*
*   already rolled away.*
She gasps.
*   The Son... had risen!*

Meditation
(a Christian tries it)

I thought I'd try it –
It's supposed to be so hard...
I sat as still as I could
and I relaxed my mind;
I opened all my doors and windows.
Then, You were there,
pouring in like the sun
with your amazing love!

	Waning Week
Gaiaday	
Sunday	In Greco-Roman myth, the Sun drives a four-horse chariot across the sky. The four horses represent the Sun on its course through the seasons.
Moonday	

The following Easter treat (found on the web) really brings the story of Jesus' gift to us home, for any child allowed to help.

Empty Tombs
Make these the day before Easter

Ingredients:

    1 cup whole pecans (in a Ziploc baggie)
    1 tsp vinegar
    3 egg whites
    1 pinch of salt
    1 cup sugar

Directions:

1. Preheat your oven to 300 degrees

2. Take the baggie of nuts and have the child(ren) beat them with a wooden spoon until they are in small pieces, and while they do it tell them about how after Jesus was arrested He, too, was beaten by Roman soldiers (John 19: 1-3)

3. Have each child smell the vinegar. Place it in a mixing bowl. Tell about how when Jesus was on the cross and thirsty, He was given blended vinegar to drink (John 19: 28-30)

4. Add egg whites to vinegar, and mention how when Jesus gave up His life (eggs are for life), He revitalized ours.

5. Put a little salt in each child's hand and have them taste it. Tell them about how Jesus' followers wept when they saw Him being taken away and remembered how they had failed

	Waning Week
Mercuray	
Vensday	Every day on Venus is longer than a year spent there, owing to how slowly it spins on its axis – and clockwise (in the opposite direction to other planets), too!
Tiwsday	

Him. Let them know this salt is for the tears shed that day.

6. Mix in the cup of sugar, and while you're doing it, tell them this sad story is also sweet, because everything Jesus went through that day He willingly did for God and us, because we are so much loved and we all belong together.

7. Beat the resulting blend with a mixer on high speed for 10-15 minutes, or until stiff peaks are formed. Tell how the color white that comes is for the purity that is restored to those who accept Jesus' gift and tell Him thank you.

8. Fold in nuts and drop by teaspoons on a wax-paper covered cookie-sheet. Mention how each mound is for the rocky tomb where Jesus' body lay.

9. Put the cookie-sheet in the oven, close the door and turn the oven off. Give each child a piece of tape for sealing the door shut. While they are doing this, talk about how Jesus' tomb was also sealed (Matt 27: 65-66) and how Jesus' followers were in despair that day (John 16: 20-22).

10. Now send everyone to bed!

11. The next morning is Easter Day. Open the oven and give everyone a cookie. Tell how when Mary visited the grave on that first Easter morning, she was amazed to see how the heavy stone sealing it had been rolled away. Take a bite. The center is hollow! Mary too, found only empty grave clothes with no body inside (Luke 24: 1-3) and an angel telling her, "He is not here, for He has risen."

	Waning Week
Jupitay	
Saturday	
Starday	For most astrologers today, planets merely represent drives in the human unconscious, or energy and how it flows along the dimensions of our experience.

Spring Moon
Aspects in Astrology

To modern astrologers, aspects matter much more than rulerships or falls. They look at how well or how poorly the energies of one planet will interact with that of another. The degrees of separation between planets are what create them. Since we observe these planets from Earth, our line of sight to one of them forms an angle with our line of sight to another one, and this angle is what defines the aspect: the type it is and the impact it will have.

There are five basic ("major") types of angle (aspect) and six additional ("minor") ones: often a variation of a basic. Major aspects describe the strongest bonds and minor aspects describe bonds that are mellower. Allowances ("orbs") are given for how rarely these aspects occur in precision such that for major aspects anything within six degrees of the angle counts, and even for the variations it counts if it falls within two or three degrees. Aspects that describe an easy connection are called "soft" and aspects that describe more blocked or challenging ones are called "hard." Among the soft angles are trines (a hundred and twenty degrees, so one-hundred-fourteen to one-hundred-twenty-six), sextiles (sixty degrees: fifty-four to sixty-six) and conjunctions (zero to six degrees); harder ones are squares (ninety degrees: eighty-four to ninety-six) and oppositions (a hundred and eighty, so one-hundred-seventy-four to one-hundred-eighty-six).

In the sky, you can see planets in conjunction sitting right next to each other. Those in opposition will be seen directly across from one another in opposite parts of the sky, and those other angles will be visible too, as described: ninety degrees apart, a hundred and twenty degrees, etc. A natal chart simply tells a person what the sky looked like when he or she was born.

Fourth Lunation

In the velvet night...

61

Dark Moon

Apple Blossoms

Apple blossoms on bare branches
still sparkle with morning dew,
dusted in jewels,
or bright rainbow hearts,
flung generously on tender petal flesh,
soft as any woman's.

I needn't be watching them to write this.
They're in my soul,
singing as I sit here,
under a fresh spring sky,
so blue...

This earth is so beautiful!
Can you see it?
What if the sky were looking at us?
Would it make you happy to see her smile?
The sun might be stroking your skin with tender fingers,
but would you ever know?

Yes, that was the other poem that wrote itself one day when by chance it was the spring equinox. This flowery lunation seemed a better place for it. And I'd like to resume my former online dreaming group's discussion of flowers, if I may.

The Discussion Resumed

"'...However, we also call the Sea Ania the Flower World.

62

	Waxing Week
Gaiaday	
Sunday	
Moonday	The second half of that earlier meditation is meant to happen one night before the moon's full (with a silver or pale blue candle). This is the time to look back on and celebrate what may have happened so far, and to notice if or in what way the dream itself (in response to these events) has changed.

Flower Moon

Everything relating to it we call sewailo, or flowery. The desert miracle of a harsh landscape suddenly bursting into bloom is a constant reminder of how the seeds and buds of Faerie already exist, hidden, in the dry mundane world, waiting for a rain of grace, of magic, of mystery-awareness to unfurl the petals, release the fragrance, cover the desert in a shivering..." Did she mean shimmering? "...rainbow of colors!" Even when I ask her, she is unsure.

"Flowers cannot be more important to the Yaqui. The elders who still speak the language greet each other with 'How is your flower?' meaning how is your soul, your mysterious core, that part of you that can perceive Faerie. Flowers are a very sacred symbol for us.'"

Twilight Knows

The mountains of Faerie glow now and then
with a soft, soft light
when the air turns thin in our world.
Deep purple shadows are cast and molded
by the sun king's gold.
When the air turns thin we hear horns, far and dim,
and songs of old.

Dream that Took Me Somewhere Flowery

You know, I also shared with them a very powerful dream I had. While it did draw out some talk of flowers, more importantly, parts of it happened in a very unusual place that reminded me of the world discussed above. In one part there was no chronological order; it felt like the nexus of a single spinning

	Waxing Week
Mercuray	To Rome, Mercury was a trickster: god of financial gain, commerce, eloquence, luck, trickery, merchants, and thieves, all relying on skills like negotiation and presentation.
Vensday	
Tiwsday	

point. I should like to share this dream here.

To begin, I'm in my mother's house, visiting with her. My stepfather cuts in, telling me the people I am traveling with are ready to go now. Is he giving us a lift? He says they are ready and heads out the door.

But I wasn't ready. I wanted to say goodbye to my mom, who was gathering up things to have me take along. There were nutritional supplements that had gotten wet, re-dried, and stuck together. Handing them to me as we talked, she went around the house looking for more things while I worked on getting what was stuck apart again. There was food with tiny ants on it; she asked me to take that outside for her. We said a hurried goodbye. I was brushing ants off the food as I walked out the door and into the garage.

There was no car in the garage. The door was open, and I went outside looking for it, but I find a street instead. I feel not sad but happy – the happiness of someone who has been freed. A lot of people are lined up on both sides of the street, expectantly waiting. I stand next to them; give the ant-laden food to the earth, the ants. The person next to me (a young, attractive Hispanic man) says there is going to be a parade. I should stick around and watch for tail-lights.

Sure enough, a horde of cars arrive soon, driving fast, coming in backwards, tail-lights first. I feel so lucky to be there, just by chance, watching for that parade. The people get out of their cars. They are all Native Americans, dressed in the ways of long ago. They leave their cars parked to walk the parade route. One man said to us in a loud voice that he was in a really lousy mood. He'd woken up that day feeling just awful, and he was still feeling REALLY BAD.

I thought he was a regular person like me, and I took him seriously. Maybe he was feeling ill, even asking for our help?

	Waxing Week
Jupitay	Jupiter rotates (spins) faster than any other planet. One rotation equals a night and a day. Jupiter's is only about 10 hours long.
Saturday	
Starday	

67

But then he began spinning around rapidly in one place, his two braids flying out in all directions to the edges of my sight, where their shadow, wide at the tips, made a shape like the blades of a Dutch windmill. He was chanting something, over and over on a pitch that rose higher and higher. We were all chanting with him and I was focusing on this with all my heart, using that chant as a prayer for his healing. I got lost in the rising pitch, the flying braids. At this instant the dream grows vague and changeful. I am in a different house – does it belong to the man with the flying braids? I am doing things, but they keep changing, from the moment I arrive.

Sometimes I do them alone and sometimes with other people, sometimes with just a few and sometimes there is a crowd; I am always engrossed in whatever activity it is. The things in my hands change the same way the people around (or not around) me do, as if all of this were happening all at once.

When I first arrive I am holding a plastic sandwich bag and putting the things I got from my mother inside it (they look like chewable vitamin pills sometimes, still stuck), but wasn't noticing (until suddenly I did) that the bag had a hole in it and they were dropping out almost as fast as I was putting them in. One kitten (who ran sulking downstairs when her brother started bothering her) was playing with these. When she ran off he kept right on playing with them.

The plastic sandwich bag in my hands turns to a paper one about the size of a lunch sack (with wrinkles, but no holes), and the nutritional snacks become decorative little bows for hanging onto a tree – silk ribbon bows framed by leafy-looking carved circles of green-painted plaster. The bows are red for a flash, then yellow; again, an orange-pink; at other times sky blue, pale teal... always one color, never mixed; now all mauve, again all burnt orange... in dizzyingly rapid succession. It is at one and the same time a plastic bag with a hole in it and a paper bag with wrinkles. Sometimes others are doing things alongside me but I

	Bright Week
Gaiaday	
Sunday	
Moonday	Luna's name comes from the verb lucere: "to shine." She was also known as Selene, meaning "Moon."

don't notice – so much shifts so fast I take nothing seriously, even when the bows of ribbon turn into paper flowers or a few real geranium pieces get mixed in.

"Hand-made flowers brings to mind that in the Yaqui community we shall soon be making tissue-paper flowers with which to decorate our churches and chapels for Easter. It is joyful work, children eagerly learning how from their elders, with long tables covered with bolts of different color tissue. We gather up the bits and snippets left over to throw as confetti for the Gloria on Easter Saturday. We also throw real flowers and petals mixed in with the confetti."

A bunch of us had made those flowers. When they changed and were no longer simple bows, the bag to hold them was suddenly made of lovely hand-made cloth of the kind one sees in Guatemala. It had grown as big as a shopping bag.

We had all been relaxing in that house doing our own things when we were invited upstairs to see a movie, in a darkened place where we would be gazing into a globe of glass. If we focused on it, it would bring the imagery for the movie up from inside itself: a miraculous thing, as those images would actually be coming from our collective minds.

I went in with everyone else, but was haunted by the idea that while I was making silk bows I may also have made a mess, so I returned to my bag to tidy up and put things back where they belonged. It wasn't supposed to take up any time at all.

I began hastily gathering up everything at hand wholesale, until (quite suddenly) I became aware of having forgotten that some of these things had been made by others. While collecting my own (olive green – not silk bows, but paper flowers), I found some of theirs in hand and remembered all at once that they weren't mine. This happened the moment the bows I'd been gathering

	Bright Week
Mercuray	Mercury still represents mental and physical dexterity, relating to education, writing, information gathering skills, journalism, even gymnastics and sleight of hand.
Vensday	
Tiwsday	

became flowers. So I thought I'd better fish them out again. This was easy at first, as each person's work had their own style and color, but again the world shifted, and though we'd all made the flowers, now they all looked alike. Did all of the work now belong to all of us? Then so did the cleaning up...

At that point I heard a lovely song and wanted to be finished. I don't understand why I didn't stop right then – I didn't like to feel I was missing out. While I was gathering the things I noticed pieces I needed would rise and come to hand without my reaching for them. I'd think of them and they'd be right there. It dawned on me that this had been happening all along: another miraculous thing. Suddenly I knew I'd heard in plain English what the two kittens were saying to each other.

"Stop doing blah blah blah..." said the little boy cat.

"I did no such thing!" said the little girl, who stormed off and ran away to the downstairs. This was the third amazing thing and it finally dawned on me how naturally I had been doing them all.

I was so excited! I gathered things up even faster (they began flying into my hands like little arrows), so eager was I to follow where the rest had gone and tell them all about it. Before I could finish, a cherished dream person with warm brown eyes (familiar, almost like an aunt) came down the stairs to me. It was wonderful to see her again. We were discussing genealogy, and she was explaining to me exactly how someone I know is related to the rest of us. I found the ability to draw up images in the glass upstairs extended to our conversation – I could project out the images inside my head so she could see them plainly, relieving us of any need for words.

I enjoyed our visit very much, even if it meant I missed still more of that movie with the beautiful song... Or could that tardiness have been a product of my patience with the kittens, who found the stuff I was putting away so fascinating. I didn't mind; the fact that we all were using the same ability to project and share images caused me to feel included, and no longer left out in any way.

	Bright Week
Jupitay	
Saturday	Saturn's rings are divided into seven sections, their size and structure due in part to gravitational influence from several moons (as a group, called "Shepherd Moons") that orbit within gaps in the main rings.
Starday	

Flower Moon

An Unusual Friend

Leewinhoo, Leewinho,
she is tall and slender, o!
Silver, gravely gentle soul,
and asks the same of you, now!
Beauty does delight her,
and the children's laughter.
A naïve curiosity
amplifies velocity:
Leewinho, Leewohin,
waking amid summers green.

The Language of Flowers

During the 1600's, Turkish concubines (who could neither read
nor write) found other ways to communicate with one another,
sending flowers as messages. This ability may have been vital;
court life could be competitive to the point of deadliness.

In 1718 the British ambassador's wife sent a letter home about
it from Constantinople, captivated by this "secret language of
flowers." Many shared her fascination. By 1819 an entire
dictionary (*Le Language des Fleurs*, written under the pen name
Madame Charlotte de la Tour) had been published. Dictionaries
(and by 1884, even books) continued to appear, the given
meanings not always in agreement with one another. The women
of Europe, South America, and even the United States gave every
sign of cherishing this new language, that allowed them to express
what could in no other way be shared under the Victorian norms
of the day. You could make known the secrets of your heart
while handing someone a bouquet. But let's hope they are using

	Waning Week
Gaiaday	
Sunday	The Sun itself is orbiting around *something*, deep in the center of the Milky Way (we have yet to discover what). It takes its sweet time: 225-250 million years per round!
Moonday	

the same dictionary as you! The floral industry has a lot of history behind it when it tells us to "say it with flowers."

For an ailing friend, these are still nice:
white (for truth) chrysanthemum – rest, cheerfulness, you're a wonderful friend
yellow rose – friendship, joy
acorns – Nordic symbol of life and immortality

Some Flower Recipes

Candied Rose Petals
Now these flowers looked lovely on my in-laws' wedding anniversary cake. They were delicious, too!

1. Gather fresh, fragrant (will be flavorful) rose petals.

2. Rinse and then dry them, removing and tossing out the bottom (bitter) white tip.

3. Separate out and beat the white of 1 egg. Add enough super-fine sugar (baker's or caster's is best) for a good consistency (you want something that will work well for use in painting).

4. Paint both sides of each petal with prepared egg wash. Sprinkle with more sugar, on both sides.

5. Rest petals on parchment paper to dry overnight.

Beautiful and delicious, the candied rose petals keep well in the freezer.

	Waning Week
Mercuray	
Vensday	The surface of Venus comes modestly veiled, in dense sulfuric acid clouds.
Tiwsday	

77

Flower Moon

Flower Water
*(will have the color, fragrance, flavor, and virtues
of the flowers it comes from)*

1. Gather fresh, flawless, fragrant flowers.

2. Put them only in distilled or filtered water, and use one part petals for every two parts of water.

3. In a double boiler (you could even use a bowl that floats on water in a pot), simmer gently on low heat. You want a light simmer – never let the mixture boil.

4. Simmer gently until all color is gone.

5. When done, remove from heat and let cool.

6. Using cheese-cloth or a fine mesh strainer, strain into a Mason jar. Store in the refrigerator, or in a cool dark place.

This water will keep for up to a year.

Fresh is best, but if unavailable, use six parts of water for every one part of dried flower.

	Waning Week
Jupitay	
Saturday	
Starday	Neptune is associated with idealism, dreams, artistry, and empathy; with dissolution, illusion, and vagueness, too.

79 |

Flower Moon
Planets in Retrograde

As seen from the earth, planets usually move through the sky in the same direction as the sun. At times, they seem to reverse their direction: an illusion called "retrograde motion," caused by their relative movement around the sun in combination with similar movement from the earth (the place from which we see it all.) Though this is an illusion, the reversal we see is real to the way we perceive things from here, so the symbolism involved still holds true.

When a planet is in retrograde, things get reversed, so much so that whatever the planet relates to is turned on its head. We re-focus energies normally facing outward so they face inward, and vice versa. Mercury does it the most, but other planets do it too. Each represents a distinct part of life.

With Jupiter, we want to expand our world, and momentum builds toward breaking free; with Saturn, we define or re-define our boundaries; and with Pluto, we spend time with our darkness, especially as reflected through our deeper relationships.

Neptune turned around is about waking up to face reality; Mars involves emotions, often unappreciated ones (like aggression or anger); and with Mercury, messages tend to be garbled so it grows harder to get things done. Yet just while things are on hold, old friends from the past are more likely to reappear, delivering messages we may also need.

With Venus, issues related to love and money stagnate, and with Uranus, it's time to get real, face our fears, and walk away from whatever is clogging up our flow.

Usually, though not the most comfortable type of event, a planet in retrograde pays off for us in the end. It's like reaming out a clogged-up sink: not fun, but sometimes necessary.

Fifth Lunation

An empty moon...

81

Dark Moon
Horse Bones

It was a dark and windy night. I had to keep my promise and take Heather home early, but with the plan of coming straight back.

"Tie your shirt-tails across your middle before you open that door. And be careful out there. I don't like the way that wind feels." We had been talking together for hours, and enjoying a potluck dinner. It all felt great – my drumming friend must be talking nonsense. How not? But suddenly upon returning to his door, I was awfully glad to be safely inside again.

We drummed and we dreamed. My feelings got stronger; I could almost see horse bones out there, reared up and lashing at the windows with skeletal hooves and naked teeth, coming from the skies and carried on the wind, the back half of the body no horse at all, only some long, thin, windy thing. We stopped our drumming and agreed emphatically that no one was to open any window under any circumstance before we resumed. Those who had to went home, those who could, stayed.

Teresa and I stayed the longest, the three of us telling one another ghost stories. So roused and enthralled were we that we barely noticed how dawn's light was stealing in. Comforted by the coming of the light, we slept at last, to wake refreshed from one of the most deliciously frightening nights I have ever known.

The Mari Lwyd

I wonder to this day what that horse of bones was. I have searched, but the only thing I can find even remotely like it is a pre-Christian custom from southwest Wales, where people

	Waxing Week
Gaiaday	The earth was formed about 4.54 billion years ago.
Sunday	
Moonday	

parade a horse's skull on a stick (draped in a sheet) through the
streets, knocking at people's doors and asking to be let in.

> *"...Midnight. Midnight. Midnight. Midnight.*
> *Hark at the hands of the clock.*
> *Out in the night the Night Mares ride;*
> *and the nightmares' hooves draw near.*
> *Dead men pummel the panes outside,*
> *and the living quake with fear..."*

(Excerpt taken from *The Ballad of the Mari Lwyd*, written by a
Welshman named Vernon Watkins, in 1941)

Within this tradition, the spirits of the dead (hungry for
warmth, feasting, and human companionship) have gathered
around their Gray Mare (the Mari Lwyd) and are marching with
her through the skies (streets), demanding entry. Folks resist
and a rhyming contest begins that goes on and on until the
house people surrender, letting the horse people come in and
share their feast. The horse folk have usually brought food to
share with the house folk anyway. No one wants the house
people to win, because the horse coming inside is considered
good luck. It happens in winter though, not in spring.

At Vern's house, too, there was feasting, and hooves of wind
drumming at the windows, with no thought for time until the
first light of morning broke our story-spell. It was beautiful: a
night we will never forget!

	Waxing Week
Mercuray	
Vensday	
Marsday	Mars has permanent ice caps at its poles. Winters are so cold they crystallize carbon dioxide, depositing it on the darkened surface as slabs of dry ice. 85

Hunted

*The majic man stalks
behind the air. Can you feel it?
The piper's flute weaves many-colored strands
of fantastic imaginings. Beware.
You will be caught. Beware.
You will dream, hum, dance stories, paint songs.*

Walpurgis Nacht

I learned of a holiday called Walpurgis Nacht (at least in Germany), that does happen near this time of year, after sundown on April 30th (May Eve). (Please see Notes, p. 266) Nearly identical holidays (each with their own names) take place in other parts of Europe. People of old would meet somewhere special to dance, light bonfires and watch for spring. Since before the 3rd Century, to Blocksburg up in the Harz range, on the highest peak of "the Brocken" (a name that brings Brecon, Bregenz, and Briga to my mind), many would come. Though conquering tribes imposed a contrary faith (p. 265), still they came. The climate of fear embraced by new faith towards old caused a new layer to be added, wherein folks in their homes (oddly, males only[11])used loud noises to chase bad spirits away. Frightened church people would keep this up all night long. Witches (strangely, all female) were the biggest fear.

"The Witches' excursion takes place on the first night in May. They ride up Blocksberg on the 1st of May, and in 12 days must dance the snow away; then Spring begins."(Jacob Grimm)

	Waxing Week
Jupitay	
Saturday	Agriculture is closely linked to seasons and to an understanding of the cyclical passage of time, so it too is associated with Saturn.
Starday	

Blessing Moon

Much is made of both May Eve and Halloween, for both mark the beginnings or ends of the two halves of the year (two halves and not four quarters being the norm both in early Celtic and Teutonic thought, apparently): one hot and one cold. May Eve was Beltane, meaning "Bright Fire," and Halloween Samhain, meaning "Summer's End." Both were seen as times when the veil between the workaday world and the intangible one seen in dreams was thinnest. At such times it was believed possible for a soul to walk out of one and into the other (and back again!)

Now why couldn't that dotty old Gray Mare have gone trotting around town on Halloween? Why ever did she have to wait 'til winter? Perhaps those companions of hers were being polite when they held back at Summer's End in hopes of kindly being invited in, but then it got too cold for the poor souls, who couldn't wait until spring. So it's lucky to share feast and fire with them? Make them wait for Beltane anyway and you might see big winter blasts come pouring through well beyond May Eve.

May Eve is called Walpurgis Nacht in Germany, after St. Walpurga (an English woman born in 770 AD who did missionary work there), because her feast day is on May 1st. She is the patron saint of coughs, storms, hydrophobia, and sailors [2]. Interestingly, that very long poem Vernon Watkins wrote repeatedly refers to the Gray Mare and her hungry souls as living in the ocean or behind the moon, when they aren't busy knocking at live peoples' doors. Walpurga's symbols are the spindle, an ear of grain, and a dog, the same as Frau Holda's (an ancient Teutonic goddess; dogs, not cats, were her familiars), so we may have here an arrangement similar to that between St.Brigit and Briga.

	Bright Week
Gaiaday	
Sunday	Like a planet, the Milky Way revolves on its own axis. It takes around two million years to complete one revolution.
Moonday	

89

Blessing Moon

Winters in Germany were long and hard, so no wonder people celebrated spring! Dancing and jumping (especially around fires) is at the center of Walpurgis Nacht tradition. Grain would grow as high as a farmer could leap. The lady of the house would leap over her broom - and old brooms were to be burned. Walpurgis Nacht fires were used to burn anything worn out over the year. Straw men (given names like illness, disease, even bad luck) could be made and burned in the fires as well. Children gathered greenery from juniper, hawthorn, ash and elder trees to hang around both house and barn (originally to give thanks and celebrate bounty, and later as protection from an evil night [3]). Extra care would be taken with the cattle; bells blessed and hung from their necks, and stable doors (triple hung with crosses) carefully locked.

Walpurgis Nacht has morphed again, and is more like Halloween today. In the Harz Mountains near the town of Thale (just one place among many) bonfires are lit, and thousands dress in witch, warlock, or devil costumes, having come from all over to party through the night. In Southern Germany, it is a night of pranks. You might find trashcans overturned or a few plants missing, come daylight. Or a garden gnome might vanish, only to send you postcards later.

Guacamole

My drumming friend loved getting together with people. We all had such fun! He used to have potlucks at his house all the time. He couldn't move very well at neck or at hip, so he depended on people pretty heavily too. He wrote a song about guacamole that makes me laugh every time I hear it because it has so many good memories for me. I think that's the kind of food the Gray Mare's companions were hungriest for: warmth, laughter, conversation,

Mercuray	
Vensday	Venus has been called Earth's sister planet. They are only 638 km different in diameter, and Venus has 81 percent of Earth's mass. She is the closest planet to the Earth. Both have a central core, a molten mantle, and a crust.
Tiwsday	

91

and good times. I'd like to share the Guacamole Song here.
Vern told me it would be okay.

The Guacamole Song

Please listen to this story,
it doesn't have much glory,
but it's true: it's guacamole;
because I love my guacamole,
it's the one and only
guacamole!

Well I made some beans and rice;
it smelled my house so nice
until I realized – no guacamole!
It sounds kind of silly,
it's good cold with chili, guacamole.

Now this hunger deep inside!
I've got to get a ride,
got to go and get some guacamole.
There's no one around,
no people and no sound – O, guacamole!

A friend knocked at my door.
"Please take me to the store.
Got to go and get some guacamole.

	Bright Week
Jupitay	
Saturday	At least 62 moons are known to orbit Saturn; 53 of these have names.
Starday	

Blessing Moon

We'll make a giant bowl,
joyful to the soul: guacamole!"

At the store at last,
got to get in fast, for guacamole.
Grocery sign said "Sorry,
please come back tommorry,
out of guacamole."

Now I'm finally home,
and I'm not alone, crying for guacamole.
Some friends heard my story,
told me not to worry,
they brought guacamole.

Now we have some beans and rice,
corn tortillas, salsa, spice,
chile rellenos, cilantro, potatoes,
lots of cheese and tomatoes...
and tons of guacamole.

Because we love our guacamole,
it's the one and only
guacamole!

Oh yes, we love our guacamole,
without it food is lonely,
guacamole!

	Waning Week
Gaiaday	
Sunday	On March 7, 321, Constantine decreed 'dies Solis' (Sunday) to be the Roman day of rest. "On the venerable day of the sun let the magistrates and people residing in cities rest, and let all workshops be closed..."
Moonday	

Blessing Moon
Making Some

After all this talk of guacamole, would you like to know how to make some of your own? There are many recipes, all simple. It's just mashed up avocados with things added to make it taste nice, like salt, or soy sauce (this from an Asian friend), or salsa, or even cilantro cut up very small. After I tasted that, there was no going back – guacamole without cilantro in it just didn't taste right anymore (though I've been known to substitute lettuce when no cilantro is to be had – it works pretty well too). One ingredient ought always to be included though – not just for flavor but to keep things from turning brown too fast, and that's lime (if you don't have any, lemon will do, but I think lime tastes best). If, however, the guacamole is to be consumed immediately, then citrus becomes an option rather than a necessity. Some people like to put chili peppers in theirs, but I never would (I don't like hot things). To show how simple it really is, here is the very first way that I used to prepare it:

1. Run your knife around the circumference of a ripe avocado, going in as far as the seed will allow; twist the two sides apart (like unscrewing a lid from a jar) and if this recipe is for one alone, put the half with the seed back in the fridge for next time. If for two, an easy way to remove the seed is to grab onto it (some say stab it with a knife, but when I tried that I found the knife can slip) and twist it out.

2. Now scoop out the contents of one half of a ripe avocado and add a dash of soy sauce.

3. Mash it all together in a small bowl. Use with salsa as a dip for chips, or enjoy on whole grain bread, perhaps with sunflower seeds, alfalfa sprouts, mustard, and cheese.

	Waning Week
Mercuray	
Vensday	Venus's surface is estimated to be around three to four hundred million years old. By comparison, the Earth's surface is about one hundred million years in age.
Tiwsday	

Blessing Moon

But that's just me. You know how to make guacamole your own way now.

It's About Time

I learned from my sociology teacher that the original Hopi language didn't have a word for time. Perhaps they already knew what a wise older friend of mine used to say: time doesn't really exist. It's an illusion! It's real because we all agree that it's real, and we measure it by things we can see - like the movement of heavenly bodies, or shadows across the land, or children growing taller - but the seriousness, the rush and the desperation we give to it is all an illusion, made real through agreements made and concepts accepted.

The first time I saw anything like this idea put into practice was as a new member of the SCA (Society for Creative Anachronisms). We were all assembling, with our costumes and our gear, waiting for Court to begin. This was supposed to happen at 9:00 a.m. I was beginning to get impatient because it was later (I don't recall how much later). When my sponsor noticed this, she advised me to relax and enjoy the sights. After all, we were on SCA Time. What was there to do but look around and visit together and enjoy ourselves?

I came across the idea again as a member of a guild that was part of the entertainment at a renaissance faire. Everyone was on Faire Time. It was okay to be flexible. We just needed to stay in good coordination with one another.

I've heard it called Indian Time at Pow-wows. There it means much the same thing: a different kind of time for a different world, offering a refreshing escape from our traditionally accepted norms.

	Waning Week
Jupitay	If Jupiter were only 80 times the size it is now, it would have thermonuclear reactions at core and have become a star instead of a planet.
Saturday	
Starday	

Blessing Moon
Stars and their Colors

Have you ever wondered why some stars are one color and some another? I have. Temperature has a lot to do with it, but composition plays a role as well, because different elements emit different wavelengths of electromagnetic radiation when heated. A star's make-up (and no two are alike) includes a few less prominent elements but is mostly hydrogen (73 percent) and helium (25 percent). The remaining 2 percent could come from anywhere, be any element that ever was.

From hottest to coolest, the star colors we usually see are deep blue, blue, blue-white, white, yellowish-white, pale yellow-orange and light orange-red. Green stars radiate right in the center of the visible light spectrum, so they emit some light in each possible color (which makes them *look* white). Our own sun sends out a lot of green light, but who sees it? Not me. And our eyes are less sensitive to purple light, so we tend to see those stars too not as they are but as the color next door: blue. A star's color can tell scientists (roughly) its temperature, elemental composition and even the atmospheric conditions surrounding it. Stars also go through an evolutionary life cycle, during which time their sizes, temperatures and colors change.

For kindly sharing this interesting information, I thank www.universetoday.com, www.livescience.com, and https://colorpsychologymeaning.com

Sixth Lunation

In the sun's arms lying...

Dark Moon
Deer (a Waking Dream)

Sometimes I can visit the people in the meditations I shared
under the second lunation, and at other times they are hard to
find. I still don't know why. I was reaching out to them one day
in the middle of summer when in a waking dream I found myself
walking through the forest in the company of many deer,
trampling the dry brush down. We were winding along the valley
floor, perhaps five abreast, kicking up dry stuff and trampling
seeds into the ground, there to rest until the rains came. Some of
us were deer, and some of us were in human form, but these
wore antlers, or deerskin, or some such. We walked through
sometimes high banks of dry stuff and left open pathways behind
us, but most of all we were simply being there with one another,
steeped each in the others' essence, both the human and the
deer.

The Legend of Herne

Around Berkshire, Herne is depicted wearing the antlers of a
stag. He is both god of the wild hunt and of the game in the
forest. He could be seen "carrying a great horn and a wooden
bow, riding a mighty black horse and accompanied by a pack of
baying hounds," but mortals who got in the way ran the risk of
being swept up in the Wild Hunt and taken away, to ride with
Herne forever more.

According to local legend, he can appear in Windsor Forest or
to the royal family to this day, if he has bad news to share like an
upcoming time of national crisis. British royals had strong
connections with the deer even in pre- Saxon times, when at
initiation the boy who would be king had to wrestle with and
defeat a real stag in the forest first. After that

	Waxing Week
Gaiaday	In an interview aired on Youtube, a spokes-woman of the Gwich'en People made an effort to explain the responsibility and protectiveness they feel towards the caribou herd they rely on. The herd lives in the Arctic National Wildlife Refuge, where a company wants to drill for oil. She said, "We carry a piece of the caribou with us in our hearts, and the caribou carry a piece of us in their hearts, too."
Sunday	
Moonday	

he had to marry the land (perhaps a priestess stood in for the land itself or channeled its essence during the ceremony, only to return to her own simple self when the ceremony was over). After he fought for the right to marry the Land and he won Her, she became his wife – someone to defend with his life, her and the family they would have together. In this case the family was all of the land and everything in it, including the humans, who honored their King as their Liege Lord.

At a music festival I attended recently I saw Beats Antique portray an amazing thing – a woman with the sweetest face. She wore a deer's antlers and may have had other deer-like features. Something about her made a deep impression on me. Was the one I saw them portraying Pretania/Brigantia herself? The King married a Deer, who is the Goddess Briga, who *is* Britain.

White Buffalo Calf Woman

I ponder over all this as I go about my days, and notice a story returning to mind time and again. It tells of White Buffalo Calf Woman, who may have some insights to share. She has tremendous meaning for traditional members of the Sioux Nation (as do buffalo themselves). When they were sovereign, buffalo was an important food. And unlike Brigantia, she isn't lost in time. This story has captured my attention. Perhaps it can illumine a few places usually dark to our memories or imaginations. I tell it here the way I remember it. It belongs to the Sioux, and it is they who will know it best.

They say when White Buffalo Calf Woman first approached the People (for she had many gifts to share) it was two young men who saw her first. She spoke to them, explaining her errand and about her desire to teach the many good and helpful things she

	Waxing Week
Mercuray	
Vensday	Venus is the third brightest object in the sky after the Sun and Moon. Long ago, Babylonians referred to her as "bright queen of the sky."
Tiwsday	

knew, but while her wisdom was evident, so was her beauty – an undeniable beauty not seen in any other. One young man listened in fascination to the wonderful things she shared, but the other was so absorbed by his desire for her that he didn't take in much. At the last he lost control of himself altogether, and seized her hand. With a gentle smile, she gave it to him. Instantly, the flesh fell from him, his bones crumbled into dust, and what had been a man fell all into a heap upon the ground. When the last grains had settled, she calmly turned back to the other man, and continued with her teachings (as told to a group of us, by one of Wallace Black Elk's students).

Some say holy things ought not to be approached in an impure way, and the lustful young man had been properly punished, as was right. But others speak of a love beyond words, filling White Buffalo Calf Woman's heart to bursting for each and every being. These say she cherished the desirous young man as deeply as she did the young student, and became his wife. For love of the man who loved her teachings, she sped up time, but for the other one, she slowed it down, living a long and beautiful life with him until that life was over. I see in my mind's eye the utter sweetness of the face that wore the antlers and find more credibility in the latter view.

Midsummer

This has long been a special time. I learn at holisticprimarycare.net that in much of Northern Europe, it's traditional to make wreaths and crowns out of the abundant herbs and flowers there and wear them on your head or put them around your animals' necks. Some would be hung in homes and workplaces. In many places these wreaths were then gathered and burned in community bonfires.

	Waxing Week
Jupitay	Orange Jupiter (ammonium hydrosulfide) is delicately robed in shimmery swirls of pale blue and white (blue ammonia and water that became ice diamonds).
Saturday	
Starday	

Here in the Southlands, real heat is coming, but summer has its own magic. A great way to enjoy it is to "take your family swimming or just turn on the sprinklers and run through, then have a bonfire or barbecue at the end of the day. Let the kids stay up late to say goodnight to the sun, or celebrate nightfall with sparklers, storytelling, and music."

One thing that makes summer special in my memories are the days we'd spend on the beach, eating chips and guzzling ice-cold things between swims, then watching the sun making paths on the water as it slipped into the west while dinner cooked over the fire. What lovely days those were!

Midsummer's full moon is a time when it's believed the fairies come out to play with humans. They love to laugh and pull pranks, "just like they did in Shakespeare's *A Midsummer Night's Dream*. Music, sweets, drink, and dancing" are things they love – or wind chimes, or bells – or anything sparkly. People often leave out food and offerings for them.

In some parts of England people believe staying up all night in the middle of a stone circle on Midsummer's Eve will have you seeing the Fae. If you get into trouble or need to escape from them, "follow a ley line; it will lead you to safety."

Candle in the Wind

I'm pretty sure I made friends with a fire fairy. Here's how it happened.

I have an enthusiastic friend who gets involved in many things. One windy evening as dusk was falling, we met as planned. There was a candlelight vigil scheduled on a street corner in the downtown area. We had everything we needed except for candles, but these were easily procured. Parking downtown is a chancy affair; we managed to

	Bright Week
Gaiaday	
Sunday	
Moonday	Here is a meditation for the full moon. Light a white candle and sit peacefully with it. Notice and name, one by one, all that is present in your life: things, people, challenges, successes, even failures. Notice your surroundings and the space you occupy in life. Labels like "good" or "bad" don't matter now. Just notice, as much as you can.

find a spot only a ten-minute walk from the site. Points of yellow light in a loosely scattered circle grew slowly larger as we approached on foot. It was quite a sight in the deepening twilight. We lit our own candles from a neighbor's flame, and immediately had to shield them from the wind, which was still strong.

In time I got so tired of shielding my candle flame that I decided not to and the wind quickly blew it out. I stood around holding a darkened candle for a while, but got tired of that, too. So I lit it again, for a change of pace. I shielded the flame, not planning to do it very long. Suddenly I remembered a ceremony I and another friend had done together a few times. It was a light-hearted thing, which I neither believed nor disbelieved in, though I appreciated the concept behind it. The intent was to offer friendship to fire elementals. I recalled a certain phrase from the ceremony: "Salamanders, led by Djinn, this candle flame you may play in," and laughingly (but silently) addressed it in my mind to the lit flame of my own candle. It was an invitation made with warmth, but jokingly, and it made me smile, because it reminded me of my other friend. And I liked the idea of being counted a friend, even by an elemental.

I brought the flame close to my face and really took it in, before dropping my hand to let the wind have it. I fully expected it to blow out soon. To my amazement, it did not. Someone must have accepted my invitation and been happily playing there, because the flame continued to dance merrily, and stayed lit for the rest of the event with no help from me.

I had gotten rather fond of the tenacious little light by the time the vigil was over and we were walking back to the car, so at first I walked very carefully. It had no trouble staying lit, even when we walked with ease. We kept it burning through the entire car ride, and even took it with us into the restaurant where we stopped to eat. I hope the salamander had fun. I sure did!

	Bright Week
Mercuray	
Vensday	Venus is the hottest planet in the solar system. The atmosphere is 96.5 percent carbon dioxide. This greenhouse effect creates average surface temperatures of 462°C (863°F).
Tiwsday	

Listen

Listen!
The drum is the heartbeat of the earth,
rhythm of all that lives.
Rainbow songs are bending over the earth;
light is dripping from them.
Clear, water-bright harmonies rise and fall,
weaving an arch of color
among the silvery stars.
Cloud songs scud across our minds,
Until, dew-sung and sun-risen,
the happening is over.

Fire's own Face

The only reason I know anything about White Buffalo Calf Woman is because long ago, one of my room-mates was a serious student of Wallace Black Elk's who made regular visits to South Dakota. Several times a month, there were sweat lodges (times of purification, with help from stone people heated in the fire), which did wonders for my lungs.

We talked of many things while steaming together in there. Among them was the nature of fire. In traditional thinking, fire is a living being, and so are stones, which is why I call them stone people here, out of respect for the many good things shared with us in there. The correct name for a Lakota (one group within the Sioux Nation) -style sweat lodge is "the Stone People's Lodge." I am happier, being respectful this way of the people who helped me.

Mostly what I remember hearing is that fire is alive, and we might enjoy looking into it sometimes, that there can be pictures in there we might like to see.

	Bright Week
Jupitay	Like the Sun, Jupiter is mostly made of hydrogen and helium (both of which are colorless) plus a few trace elements.
Saturday	
Starday	

We were advised that if we were friendly and kind fire might decide s/he likes us, and show us its own face. This was long ago in my life, but I loved the concept and stored it up in my heart.

A few years back we went on a camping trip through the Pacific Northwest, visiting family, going to that music festival, spending time with friends, and having a lot of quiet time in between. A short drive ending at a lovely campsite held one of the richest highlights of the trip for me. The owner drove around in his cart with firewood to sell. Neither one of us had much confidence with fire, but we decided to try our hands at building one anyway. James bought goopy stuff to make the wood catch more easily and tried for a little while, but soon had me taking over. I drew on what I'd learned but had not had much chance to use yet – starting with small dry things with access to a succession of gradually bigger ones stored close by. The fact that goop had gotten smeared on some of the stuff already no doubt helped.

I'm someone who talks out loud sometimes even with no visible person at hand to hear. I'd long held the idea in mind about fire being alive, and how sociable it is. It was just easier for me, talking to the fire while seeking to build it a good place and draw it out from wherever else it lives so it could be bright and pretty for us at our quiet campsite there. We soon had our warm fire, crackling merrily. I stuck close by to keep it fed, and watched the pictures forming in the coals. One that lasted a particularly long time showed a female figure sitting cross-legged with her back to us and a long braid going down. *I* had a long braid down my back; *I* sit cross-legged a lot: the picture looked a lot like *me!* In time, a new one formed. This time it was more masculine, showing a stumpy figure and a face with large round eyes. Once or twice the flame flickered in a certain way. It was very subtle: did one eye just wink? Then it changed again, and I saw another face. The eyes were crinkled up

	Waning Week
Gaiaday	
Sunday	A mosaic floor in Hamat Tiberias presents King David as Helios surrounded by a ring with the signs of the zodiac. Such figures also appear in several of the few surviving schemes of decoration in Late Antique synagogues like Beth Alpha, Husefa, and Naaran.
Moonday	

as if in laughter, and the mouth curved as if it were lauging too! I really feel I'd made a friend again – I think I built a fire that liked me.

Midsummer Fires

An excerpt from Hutton's *The Stations of the Sun: a History of the Ritual Year in Britain* tells that of old, Midsummer was a time to honor the space between earth and sky. Well into the 19th Century, people were known to light festive fires on hilltops all over Europe, and in North-Western Africa as well.

In England farmers used to light a fire on their land (perhaps some still do), "and people holding torches and lanterns would wander around from one bonfire to another. If you jumped over a bonfire, you were guaranteed to have good luck for the coming year." Traditionally, "cinders from the fires were scattered over the fields as an 'offering' to protect the crops." That's a practical idea. Perhaps the cinders had nutritional value for those crops.

Navaho Bread

These days, cooking over a fire (especially one you build yourself, out of materials you gather from land you are on) isn't done very often. So it was quite a treat for me when I was visiting those nice folks in the Navaho Nation to help one of the daughters gather fire materials, and to help shape the dough so it could be cooked over the coals. It was very versatile dough. They called it **Fire Bread** when it was cooked like that. Then again, it could be baked in an oven; cooked that way, it made lovely **breakfast biscuits**. I asked the eldest sister for the recipe. It's simple but

	Waning Week
Mercuray	Other possible roots for Mercury's name are the Latin words 'merx' (merchandise), 'mercari' (to trade), and 'merces' (wages).
Vensday	
Tiwsday	

practical. These are the **Ingredients:**
Flour, salt, baking powder, powdered milk, and water...

1. Take flour, with salt added to taste, a little powdered milk
   for softness, and a smidgen of baking powder.

2. Blend a little at a time with water until it has a smooth and
   elastic consistency.

3. Knead it a lot for fire bread so it can be shaped and
   stretched big like a flour tortilla – you'll be cooking this on a
   rack over coals. Or...

4. For the biscuits, just gently fold it together. Put some dough
   on an oiled baking sheet and bake at 350 for 20 minutes –
   they should rise quickly.

When fried in a pan with oil the dough you stretched like a
tortilla becomes **Fry Bread:** a tasty base for those highly sought
after Indian Tacos (called Navaho Tacos inside the Navaho
Nation) sold at pow-wows across the land. If you've ever been to
one then you've seen them, too – heaped high with beans and
meat, cheese, tomatoes and onions, sour cream or guacamole,
chilis or olives: any good thing that can be thought of just might
be included. Each place makes them differently (and almost
always delicious). They make for a hearty and satisfying meal.
   I've also bought them dusted with powdered sugar or laced
with honey. One place even sold them dotted in cream and
buried under strawberries. They were wonderful made sweet that
way, too. Whatever else this dough might be, limited it isn't.

	Waning Week
Jupitay	A temple to Jupiter was built in Rome topped with statues of four horses, drawing the same type of chariot as was driven by the Sun but in this case, Jupiter was driving.
Saturday	
Starday	

Summer Moon

The Suns' Magnetic Field

The Sun has a very large and complex magnetic field, created when the positively charged ions and negatively charged electrons forming the Sun's plasma vibrate very quickly (in response to extreme heat generated by the fusion happening at its core). This field gets looped into twists and turns through the movement of the Sun around its axis. The Sun's gaseous state complicates this looping further, for everything moves faster near its equator (doing a complete rotation once every 24 days) than near its poles (once every 35 days). We can tell by watching sunspots (cooler areas) move across the Sun. So the difference between poles and equator in relation to solar spin results in still more twisting and stretching of the field formed there. Particles move so quickly and turn so tightly that spin-off (which forms the solar wind) is inevitable, and this spin-off-built wind creates its *own* field, complicating the whole mess still more.

Close to the surface everything is twisted and complex, but deeper down, some general trends emerge. The magnetic field is stronger nearer the poles and weaker closer to the equator. However, even there, it is still around 100 times stronger than Earth's. The Sun's magnetic field is so large it influences the motion of charged particles well beyond the orbits of all known planets.

Looking Up

Did you know – the sky is always aglow.
Somewhere the dawn is coming, or the sunset is going,
lovely as the rainbow. And look!
There's happiness – running loose, under the stars!
We have but to see and then claim it for ours.

Seventh Lunation

Deep in stillness, scrying...

121

Dark Moon

The Moonhutch (a Process)
"She secluded herself, in imitation of the moon."

A practice I find fascinating has evolved from one woman (Ruth Cox), her struggle with connecting to her creative source, and her search for a means of maintaining that connection. This had proven necessary for her, as her lack of such was affecting her on every level including the physical, and manifesting most painfully during her menstrual times (when, according to a number of non-westernized cultures, women are most sensitive to and most aware of subtleties within themselves and their environs). She built the foundation for her search around these times – the pain itself would remind and motivate her to continue her practice.

Through her discoveries she found such relief, balance, and renewed harmony that she created a website called "The Moonhutch" (many years ago - it may not exist anymore) in order to share her practice with the rest of us. It had such an impact on me I took notes, in the form of a poem:

Moon time – when the veil between
the conscious and unconscious is most thin –
so good, just tap in. Collage is an open door.
Light a candle – ritual space –
or music gives a lead – just breathe.
From gathered magazines the pages turn,
we search for images that burn –
love, hate, everything we feel: these will lead us to what's real.
The page is black, and images go stumbling forth,
each to find their place, their worth.
The divine right to create
has us whole again, feels great.

	Waxing Week
Gaiaday	
Sunday	
Moonday	1)Moonlight has an impact on human physiology, and those who sleep in places that are open to the moon's changing light discover, over time, how their personal (and more visibly, their menstrual) cycles recalibrate to harmonize with those of the moon. 2)In earlier days, during pregnancy these lunar cycles with their phases helped people track the passage of gestation. 1) Concept comes from Louise Lacey's book, "Lunaception." 2) Material is from a Wikipedia article 123

Hot Moon

Menstruation, sexuality, emotion, vitality,
creation, life force, delight, remorse –
all parts of whole: divine.
Insight, perception, awareness, vision
bring wisdom, help realign.
The natural rhythms of blood time
is pulse of the Mother through time.
They used to know
the spirit is embodied in the blood –
gifts these are, of love.

Finding a way to let the unconscious surface is one of the primary intentions of this practice. Ruth found that by stepping outside of her ordinary life and going into a quiet space, made special with music and candles where on black construction paper she formed collages of pictures taken from magazines, all manner of insights flowed out to her perception. By paying attention month by month, she was able to change what needed to be changed in her home, her family, and her life; and in time, whenever the time for bleeding would come, it could do so without pain.

Walks on the Beach

Bob and I were walking on the beach one day, as we had done many times before – all part of a plan we had in place to walk the entire San Diego coastline from the Mexican border to the Orange County line.

We'd already spent many delicious hours this way, catching random buses to various parts of the coast and walking the shoreline

	Waxing Week
Mercuray	
Vensday	On Venus, atmospheric pressure is 92 times stronger than on Earth: enough to crush any small asteroids entering its atmosphere. So no craters or blemishes on *this* lady's skin!
Tiwsday	

until we'd had our fill for the day, then catching the nearest bus home. We kept track of where we'd been to avoid repeating ourselves, and were slowly marking off the places we'd already seen. Every day was special, but this one was different somehow. Bob noticed it first.

As we were walking along the shoreline, he pointed out a little finger of a wave. Together, we watched it move far up the sand in a long, skinny path to wet down part of the blanket belonging to a couple that had been sunning there. Apparently this wasn't the first time it had done something like that – he must have already been noticing a few odd things, to be pointing it out to me and predicting the mischief before it took place. They promptly moved a bit higher up the shore, but then another little finger formed – it appeared to be aiming for a different group of people, close by.

Right about then, Bob found a ping-pong ball that had been lost in the sand. He tossed it into the water to see what would happen. The people in that second group were spared. The little finger dropped back, and waves took the light little ball instead. Then they brought it back again: a few inches to the right of my right foot (Bob was standing at my left).

"Oh, look, Bob! Do you think it wants to play?" Well, he poo-pooed *that* idea, but while we were talking about it the water took the ball back out and set it down, next to him this time (a few inches to the left of his left foot). He started reaching for the ball himself, but I stepped in and tossed it before he could get it. It had come to me first. I didn't want to miss out.

As it happens, I needn't have been concerned. There was plenty of play to be had this day. The wave brought my tossed ball back and set it diplomatically down into the space halfway between my left foot and his right one. We took turns tossing the ball into the water, which always brought it back to us and always settled it in exactly the same spot.

Now Bob is scientifically minded. He tests things out, tries

	Waxing Week
Jupitay	
Saturday	
Starday	Some astrologers feel planets have an influence all their own, perhaps through gravity or some undiscovered source; to others they merely mirror universal patterns that include the ebb and flow of human impulses, repeating everywhere in fractal-like ways.

doing things different ways, and pays close attention to whatever goes on around him. We'd walk a few steps (and sometimes many more than that) up or down the shoreline, toss the ball this way and that... I think he was looking for differently running currents to drop the ball into, or testing to see if there wasn't a hidden one where we'd first been standing; but wherever we went, between our legs was where the ball would land, without fail. Once, Bob threw it out into deeper water, and I thought it had been lost for sure. It came back to us even then, though it needed an extra wave or two to get the job done that time. The waves were as constant as my grandfather's black Labs, and far more precise!

Suddenly it was time to go. I threw the ball higher up the beach and ran back to thank the water for the game. I stopped paying attention after that, but not Bob. Apparently the ocean had not yet tired of noticing ordinary people and trying out different ways to play with them. He says the other couple the water had been aiming for earlier got a wetting next – or some of their blanket did!

It was beautiful, friendly 70-degree water. I'm so glad we were there and were included in the fun. I'd have missed it for sure if I'd been out there on my own. I hadn't been paying attention.

The Gray Strand

I walked down by the gray strand
to seek the shadowed lee –
strange creatures, maybe, in the sand,
or treasures washed by the sea.
The misty, hissing, roaring sound
of great waters soaked through my mind;
bright dreams fell dripping on the ground
of my soul, for me to find,

	Bright Week
Gaiaday	About 71 percent of the earth's surface is under water, but only three percent of water anywhere on earth is fresh, and two thirds of that fresh water is frozen.
Sunday	
Moonday	

*and strange sea songs came floating down
through the air, like horns on the wind;
so I soaked them up, like an old sea-sponge,
at night to dream of them.*

Sea Mists

*Wraithlike,
spirit-like,
born of the sea;
of sea-foam,
of salt spray,
wild and free.*

The Nature of Water

Not long ago, I had a chance to learn a little more from the Lakota (or at least the Standing Rock Sioux) Nation – we all did. It was online.

Some of them were trying to protect the water they relied on for life and health, which was in danger of being poisoned by a pipeline they didn't want cutting through their land. So they went online and shared their struggle, hoping for some help.

I wanted to help, but couldn't add much to a GoFundMe account just then, so I sent out a few prayers, and listened to everything of theirs I could find. They said something very beautiful about water. They said when you get ready to pray, if you can, go to a place that's near water, that the water will hear you, and carry your prayers along wherever it goes. I loved the way they put it, but I may not be remembering it well. "My granddaughter is praying..." is how they said the water's thoughts would go. I sure wish I could remember

	Bright Week
Mercuray	
Vensday	Scientists believe Venus once possessed large amounts of water, even oceans. But this water boiled off long ago; the planet is far too hot to sustain life now.
Tiwsday	

more. I believe they said the water would be pleased, and carry the messages of our caring along, and do whatever it could to help.

Native Americans are not the only ones who have noticed something special about water. I learn from Sadh Guru out at Isha Foundation that the traditional people of India, too, take their water very seriously. They never drink it straight from the tap. Instead, they put it in a beautiful container and keep it in a lovely peaceful place with a flower resting on the lid. In their eyes, it has taken in all the violence of its journey to them – the fast movement, the sharp twists and turns of the confining pipe – and they don't want to take that violence into themselves. So they welcome the water, let it rest for a day. They greet it before taking it into themselves. They want to be friends with it.

There are people (Jaques Beneveniste for one) who have been interested in exploring ideas like these, but unfortunately such efforts have not been well received by the scientific community. Despite being a candidate for the Nobel Prize at the time, the day he published a paper in a respected magazine addressing his initial inquiries into the matter and the results that were obtained was the day the process was begun by which he was stripped of credibility and lost access to important resources. The story of this stripping has two sides, each quite credible on its own but differing sharply one from the other. If you choose to look deeper, try to get a look at both before drawing any conclusions.

From Luc Montaignier (a molecular biologist who won the Nobel Prize for discovering the existence of the HIV molecule) I learn that India is wise to handle water in the way they do. Apparently water molecules can link together almost as if they were holding hands, and form long chains that sometimes form into circles (referred to scientifically as "domains of coherence"). Nothing can cross such a line, once it has formed, except for

	Bright Week
Jupitay	
Saturday	Saturn is visibly flat. Its diameter at the equator and poles varies by almost 10 percent: the result of rapid rotation in a fluid state. The other three gas planets are oblate too, but not as much so. Saturn is the least dense; its specific gravity is less than that of water.
Starday	

electromagnetic signals. These can cross over, and therein remain preserved for some time. So it makes sense that whatever radiates out electromagnetically can have an impact on water, leave its signature preserved in the medium, and from within that space have an impact (albeit perhaps quite subtly) on whatever organisms take that water in (for more: Back Pages, pp. 266-7).

Electromagnetic signals are very sensitive and not well understood. There may be many factors affecting them of which we as yet know nothing, but this one aspect at least we comprehend. Seen in this light, the possibility of water being responsive to qualities surrounding it makes scientific sense. But how *do* we explain an ocean turning itself into a Labrador Retriever for the day? That I'll never know!

Missed

The rain
was singing that night.
It sang silver
all over the streets,
down the leaves of the trees,
all over the sky –
in beautiful harmony with the moon.
Ah, the rain!
To walk among windy trees...
How they would be glowing,
leaves uplifted
toward the sky,
dancing in the wind!

	Waning Week
Gaiaday	
Sunday	
Moonday	The moon is large enough for its gravity to affect the earth, stabilizing its orbit and producing the regular ebb and flow of the tides. Its cyclical nature has proven helpful to humans as well, aligning us as we move through life's natural rhythms, deal with change, and process even the difficult emotions.

Hot Moon

Mist

*Mist falls so softly, blotting out the land,
turning mountains into faerie castles.*

St. Swithin's Day

A 9th Century event generated a legend, which generated a
holiday, which we've forgotten all about by now, but they
celebrated it happily enough in medieval times.

A man by the name of Swithin lived in England and was the
bishop of Winchester. He told the people before he died he
wanted a simple burial, in a plain grave out in the churchyard.
They honored his request at first, but a hundred years later, they
wanted a memorial for him that was much more ornate. So they
built it, dug him up, and put his bones in it.

Very soon, the storm clouds gathered, the thunder rolled, and
the rain came pouring down. Did it rain without ceasing for forty
days and forty nights? It rained long enough to get their attention,
because soon they were calling it "the bishop's tears." So they
moved his bones again – to a nice, quiet spot inside the cathedral.
The sun popped back out, and they figured St. Swithin was
content.

"Wet or dry, just so, forty more days will fly." "If it be fair, for
forty days t'will rain na mair, if it does rain, for forty days it will
remain." St. Swithin's Day in England was like Groundhog Day
is in the U.S. Yet it was more, for the day had certain customs
and foods attached.

Folks would sit at table in groups of five, where large, polished
apples had been cut crosswise to reveal the star pattern within.

	Waning Week
Mercuray	
Vensday	Venus has no seasons, for her axis is not tilted. The slow movement of solar winds across the surface keep temperatures between night and day from varying much as well.
Tiwsday	

Hot Moon

This five-sided design was believed to represent the seasonal round in its entirety. Some halves were then dipped in a bowl of salty water, and everyone would eat a bit in memory of St Swithin's tears. The others were for dipping in a bowl of whipped cream, tinted gold with the aid of saffron or marigold or dandelion petals. These would be eaten in memory of his joy, but the golden sun color symbolized timelessness, too. People would eat the two together and make wishes for a long life – a life lived in balance, a life filled with right choices.

Games would be played on this day taken from all around the various seasons, like bobbing for apples, and playing circle games around something very like a May Pole. Races would be run, and there were other treats too (like plum or current bread), bringing the baking fests of Lammas to mind. The day was all about celebrating summer's bounty, and they did say that fruits and vegetables watered by St. Swithin's rain tasted the best of all. I learned of these things and more from M. P. Cosman's *Medieval Holidays and Festivals* (pp 70-72).

A Summer Drink

I enjoy summer's bounty too, and love to visit Farmers' Markets. One day I discovered Chocolate Mint. It doesn't taste or smell like chocolate at all, but is every bit as strong a peppermint as that found in a peppermint patty. It helps me feel cool even when I'm hot, and blends nicely in a summer drink I like to freeze and then take along on the hottest days. It's pretty simple to make. You just

- Place several sprigs of Chocolate Mint in a teapot.
- Add a glass of boiling water.

138

	Waning Week
Jupitay	
Saturday	Though Saturn is mostly hydrogen and helium, it is more liquid than gas (hydrogen becomes liquid when density is above 0.01 g/cm^3). Temperature, pressure, and density all rise steadily toward the core, causing Saturn's hydrogen to become a metal at the deepest layers
Starday	

Hot Moon

- Let steep and then cool.
- Add a glass or two of apple juice, and
- pour the blend into a water-bottle.
- Chill or freeze the results.

Telling Time by the Stars

I hear the early Celts were night owls and tended to stay up quite late, but that if they wanted to know what time it was, all they needed to do was take a look at the stars. I was mystified by this. I could understand getting a pretty good sense for how many hours of daylight remained by looking at the sun and seeing how the shadows lay across the land, but how anyone could predict the number of hours before sunrise was beyond me. Apparently, it's no strange thing; the sun's not the only thing that rises and sets. The moon might move around too fast to be much help, but constellations go more slowly. They rise and set too.

Where I live, Orion is bright, unique and large. It would make a lovely time-piece for fall. In summer Scorpius is too bright to miss, and is scattered over half the sky. I do prefer the old Hawaiian concept I heard for it online, though. In their islands, Scorpius was Maui's Fish Hook. On Java in Indonesia, it was known as "the Brooded Swan," or "the Leaning Coconut Tree." This I learned from https://brickthology.com › category › javanese. Now I don't go out much at night in winter or in spring, but if I did, no doubt I could name you some fine celestial time-pieces for those parts of the year. I'm pretty sure that you could, too, no matter where you happen to live; you've only to go out at night and look.

Eighth Lunation

That silent tune...

141

Dark Moon
Mama Lion

As a young girl, I had a dream. It fascinated me so much that off and on – for over 30 years! – I'd find myself pondering it when least expected. I dreamed someone came and lifted me up, carrying me easily in her arms. We must have passed through someone's living room, for while carrying me she walked right into the TV set (!) and then out the backside.

There were lots of tubes and such on the backside, in a long, carpeted hallway filled with such things; but farther on, down that hallway and out the door, there was a completely different world, of deep woods and pale pink sky. She took me to a tall white tower, upon which she set me down. Soon we were flying through the air with me astride her back, and landing on a floating dock over dark blue water...

The things that happened in that dream were intriguing enough, but what captured and kept my attention all those years was the person herself – the one carrying me to all those places. I had an overwhelming curiosity about who she was and what she was doing in my dream. She didn't feel like anyone I could imagine. To me she felt powerfully real. She was very tall. Her long legs and arms moved with unspeakable grace and freedom. Other than her waist-length red hair, she wore no clothing of any kind. Her skin was so milky white, it appeared to glow. But it was her eyes that fascinated me most. They were a deep lavender-violet, and though warm, utterly unconcerned with anything not pertinent to the purpose at hand (whatever that might be).

I would often ponder on who she may have been. While engrossed in this an image passed through my mind one day so lovely I wanted to paint it. In seared oranges, browns, and

	Waxing Week
Gaiaday	
Sunday	Light from the sun looks white because it is every color mixed together. It takes an average of about 8 minutes for light from the sun to reach our eyes, but it took millions of years for that light to emerge from the sun's core out to its surface.
Moonday	

golds on a dry grassy plain she walked, surrounded by a lion and his two lionesses. Though still completely naked, what with the way one lioness was strolling between her person and our view and the way her long hair flowed over and around her, the whole scene could have been painted quite candidly and still been in very good taste.

Intrigued but not satisfied, that one glimpse given made me wonder that much more: Who *is* this lady of the lions? I must have asked myself this question at least a thousand times. I never found what feels like a definitive answer.

One known as Astarte (Ashtoreth) in the west and as Ishtar (goddess of the morning and evening star) in Babylon *was* associated with lions. But then, Asherah (a West Semitic mother goddess), too, has at times been called "the Lion Lady," though her principal epithet was "She Who Walks on the Sea," or less commonly, the Qadesh ("Holiness"). And these two are not unique in their association with lions. In fact, the lady of the lions is "an image that extends across time for more than 6,000 years and across a wide geographic range: from Minoan Crete in the west and Anatolian Turkey in the north to Mesopotamian Sumer (Babylon, or modern Iraq) in the east." Mother Asherah "would be depicted with lions through the apex of Canaanite civilization and beyond, to the days when Israel ruled that area."

At root, it seems, we've all known Mama Lion. Some of the primary associations with lions are "strength, power and protection. They often appear in positions suggesting they are guarding something important." They guarded the thrones of goddesses, "perhaps even before the invention of kingship. Long, long ago, back in the Neolithic towns of Anatolia" (believed to be the home of the world's first farmers), they guarded her throne. "In other words, lions have been the companions and perhaps the guardians of goddesses since the beginnings of what we call

	Waxing Week
Mercuray	
Vensday	
Tiwsday	Because Mars is outside of Earth's orbit, it's the first and nearest planet to not set at about the same time as the Sun.

Western Civilization..." And the concept found there seems to have migrated throughout the Old World (www.britannica.com/topic/Asherah-Semitic-goddess).

So what was she doing in my dream? *I* had never heard of her – I was a good Christian girl. Actually, I was a very lonely little girl. My mother and I had been quite close when I was tiny. We were best friends. She had been my *first* friend, and then one day we lost one another, through a misunderstanding that never cleared up until I was grown and gone. Oh how I longed for her return! I never understood what happened. I was afraid there was something wrong with me and I hid from the world in books.

Could all the longing for Mother have called out the *Great* Mother? I'll never know. But isn't it strange that so pale a being should be found in the company of such sun-drenched beasts? The red hair and violet eyes I've heard mentioned before once only – on the mother of a girl to whom I strongly related in a favorite book. The two were very close. Could she have borrowed the coloring to infuse a deeper sense of motherliness in the rays of her presence? There's no way to know. But those mythic images and unanswerable questions *did* bring comfort – many a time, over all those years.

Lion Weave

"In the lion,
there's a heart breaking by the fire..."
(Inarticulate, it cries.
With no mobility, it flies –)

	Waxing Week
Jupitay	
Saturday	
	Astrology's famous "Saturn Return" happens when Saturn occupies the same sign it did at a person's birth. This is when we reap what we've sown, and valuable but hard lessons return to view. If we haven't mastered them yet, we do now.
Starday	

"and in the fire,
there's a heat enough to cage around your soul."
(dreams wings, sighs seas;)
"...but in the fire,
there's a heat to melt the cage around your soul."
(sometimes sings, sometimes flees).

Quoted portions are from Alex Ebert's "In the Lion"
(shared by permission).

Lammas

No significant holidays survive for this time of year, but in Merry Olde they made a party out of Lammas (August's first days). The wheat was in. They were celebrating bread. Bakers would "teach geometry" by baking bread of every shape; they'd make animals or castles and such, too, for use in reenacting scenes.

Young couples married a year and a day could compete for a prize. They had to prove they'd not regretted their union ever. Several pairs would stand before a "jury," with a "judge" presiding. The "judge" would present absurd, amusing scenarios to each wife or husband, asking "was there no joylessness? no jealousy? no jangling?" The couples now had to show (creatively and imaginatively) how each "calamity" proved a blessing or an opportunity. The most entertaining pair of liars won. The prize was half a pig.

Out in the fields, the gates would be swung open. Every kind of animal was welcome. All ate, as freely as they wished. At parties, each person leaving would be given a small loaf to take along.

	Bright Week
Gaiaday	In the past five decades, development experts estimate that as many as 2 million people have moved to reclaimed lands that had formerly been parts of the Sinai and Sahara deserts. Such reclaimed plots now account for almost 25 percent of Egypt's 8 million acres under cultivation.
Sunday	
Moonday	

Three quarters of it would be meant for him or her, but s/he was supposed to make crumbs of the last quarter to scatter on an early morning and share with birds.

Lammas/Lughnasa was the old Celtic name for this holiday. It was likely celebrated one way or another in all the different Celtic lands. This too I learned from Madeleine Pelner Cosman's *Medieval Holidays and Festivals* (pp 73-76).

"...Since Sliced Bread..."

At Reuters on the web, I learn that starchy bits have been found on 30,000-year-old grinding stones. This was in Paleolithic Europe and the bits in question came from roots, but in the Americas, too, starches played an important role.

The folks who are the original inhabitants of the land where I lived made theirs out of acorn meal in long-ago days, (and still do, at special times set aside for such things). They called it Weewish. I went to school with some of them. A friend I had who is part of that culture made a joke about it once and said "We wish this was something else." She may not have said that very thing, but she did say something like it. So bread in some form or another has been with us for a very, very long time.

And sliced bread – how long have we had that, do you imagine? The answer: less than a hundred years. At www.history.com I learn "the first commercial, automatically sliced loaves were produced on July 6th, 1928, in Chillicothe, Missouri..."

Blackberry "Syrup" and Blackberry-Apple "Jam"

I used to make a point of gathering the wild blackberries that grew a half a mile away from me, for jam. One day I got tired of

	Bright Week
Mercuray	Mercury's name may be related to the Proto-Indo-European root merĝ: 'boundary, border' (Old English 'mearc', Old Norse 'mark' and Latin 'margō').
Vensday	
Tiwsday	

buying pectin from the store and thought I'd try to make my own, by extracting that which occurred naturally in certain apples (you boil them down, or so I had read).

I tried a few times. The "jam" never would jell very quickly, but I jarred it anyway and put it aside for later, just as I did with a concoction of berries and sugar I meant to use as syrup, since I'd used no pectin there, either. That apple-berry blend never jelled at all (the flavor was incredible though); my pectin experiment, it seemed, was a flop. I liked it so much I made lots more anyway, and I liked the blackberry syrup too, but it took me a while to get around to using it.

By the time I did a surprise was waiting: the stuff had turned into jam during the time it spent sitting around! Apparently enough not-quite-ripe berries had gotten into the mix to do some decent jelling after all. Perhaps natural pectin can be found in *many* things, if they are sour enough.

Lady Sparrow
Shamelessly filked from "Lady Bransle"

Oh, she will preen her buddies in spring,
and dance among the flowers;
in summer's heat her trilling is sweet –
she sings in leafy bowers;
from 'mong the canes she seeks scattered grains,
or fruits of fall not foundered;
brown and gold in wintery cold,
she fluffs her feathers 'round her...

	Bright Week
Jupitay	The word Jupiter comes from Proto-italic *djous/dieus 'day, sky' + *patēr 'father' and means "Sky Father." Zeus's name is drawn from that 'day' or 'sky' word, too.
Saturday	
Starday	

About Birds

Birds, I've come to understand, have a very special connection with music – much deeper than most of us would suspect. Do you wonder how I grew to notice this? Not through my own observations! But I've been blessed with very observant friends. Bob, in particular, made a steady habit of taking regular breaks from whatever project he happened to be working on, sitting back and simply paying attention to everything around him. He'd do this for ten or fifteen minutes at a time. Apparently, he found it the most satisfying way to rest.

While we were living in the mountains, I began a habit of regularly practicing every tune I knew on my pennywhistle. I had been quite sporadic about taking time to do this while I lived in town, but out there, the bus would only pass by once every three hours, and its nearest stop was six miles away. Whenever the car would break down (and sometimes it stayed that way for quite a long time) we relied on this bus. It was very hard to predict just how long a wait we might be facing whenever we wanted to go into town, so I always took my pennywhistle along, for entertainment.

The more regularly I played it, the more I came to enjoy it. I grew to be quite comfortable with my instrument in a way that was obvious to anyone who happened to be listening nearby, for I'd come to enjoy it so much I even played where the waits were shorter, at bus stops in town. This is how I came to adopt a habit of regular practice, which I enjoyed very much.

Well, I was deeply engrossed in practicing one morning, not paying attention to anything except the music. I was playing outdoors – the weather was beautiful, and the surroundings, as always, lovely. Bob was working on a project outdoors and sitting back during one of his breaks. What he noticed happening amazes me. Apparently the birds were enjoying my

	Waning Week
Gaiaday	
Sunday	
Moonday	This meditation is for the Crone Moon, when the moon is in its waning phase. The time for it is one night after the full. Here, we name things we have that no longer serve, or we no longer want. Clothes that don't fit, limitations and fears we're tired of tolerating, weeds, and plant parts we prune away all fit nicely here. This is an especially good time for letting go of grief and saying goodbye to what has been lost already. Use maybe a black candle.

music, too. They were singing the tunes back to one another!

Now I myself have noticed something about my neighbor Hazel's birds (she has cockatoos, and for a time, I helped her care for them). Not only do they use human words – they use them intelligently. A few of them were quite sociable and would almost always come right over whenever I'd come around – to clean their water bowls or do whatever else was necessary. I wore an old dilapidated straw hat to keep off the sun, for it was summer at the time, and the hat helped. Those large beaks of theirs made me nervous. But the hat was already falling apart; I'd let them tear off little bits of it as a distraction while I worked. They loved to do it – loved playing with the straw. Well, on one of those occasions a bird accidently dropped a particularly appealing piece between the runs. I knew she'd been spending too much time with humans when she reflexively said "Oh, shit!"

I'm not the only one who has noticed interesting things about their birds. My mother's friend has parrots and she lets them fly free in the house. Mother told us this story at a party when we were playing "Two Truths and a Lie." A number of us picked this story out as a lie, even though it is true.

Her friend enjoys music and was taking piano lessons. One of her birds was sitting on her shoulder while she practiced a particularly difficult piece one day, and it said to her "too fast." My mother's friend ignored the advice, taking it for the unconsidered ramblings of a bird that was merely playing with sounds. However, when she played the piece for her music teacher she was told the same thing! So she went home, practiced the piece some more, and this time she did it more slowly. Her parrot was close at hand again (perhaps it enjoys music as much as it enjoys her), and while she was playing, the bird said to her, "better." I'm told there are many equally fascinating and true stories she could tell us about her bird friends.

Teresa told me of an adventure had with wild song birds,

	Waning Week
Mercuray	
Vensday	
Tiwsday	Picus (the woodpecker) is sacred to Mars because of its beak, so strong it can peck at oaks until it reaches the inmost core. It was thought to hold Mars' power and ward off harm. In mythology Picus was human and could foresee the future, even after he was changed into a woodpecker. In one tradition, he was Mars' son. The Italic Picenes were said to take their name from the picus, which they used as their animal guide during regularly undertaken migrations embarked upon as a rite of Mars. 157

which begins with her so much in need of both sleep and haste that she would pre-set the alarm on her cell phone to repeat ad infinitum until manually turned off. Birdcalls were the wake-up song of choice. Racing through the morning routine with eyes on a bus that must not be missed, she rarely bothered to turn off the alarm until she was safely waiting at the stop. Yet she was never so rushed as to not notice the hush that falls on places loud with birdsong when someone new arrives – almost as if the birds were checking the newbies out. Often, she would mentally greet them as she went by and think nothing of it; though she did get a kick out of how many of the birds would sing back to her the sounds her cell phone alarm made. This put me in mind of Bernd Heinrich's *Ravens in Winter*, wherein he posits that ravens name themselves (p. 249):

'Vocal mimicry is another source of call variation...individual-specific calls (often mimicry of other sounds, such as a dog's bark or a grouse's call) served the ravens as "names." For example, if a raven's mate was removed, the distraught partner called for it by mimicking the missing mate's individually-specific call.' Putting two and two together, we think the birds had been greeting my friend by what passed for them as *her* name, since throughout her time near them she (or her phone) had continuously been announcing it.

Based on things I've shared with you here, I see birds differently from the way I once did: as virtuosos of sound who truly love music, and understand more than we credit them to do.

Casting

Come! Cast your nets through the stars,
and see what might be.
Find what might be yours in the misty sea.
Haste! Seek the golden sails in the sky above,
and follow the tales that tell of love!

	Waning Week
Jupitay	
Saturday	Saturn's rings remained unique in the known solar system until 1977, when very faint rings were discovered around Uranus and soon after around Jupiter and Neptune, all gas giants like Saturn.
Starday	

Berry Moon
Why Not

Why not evolve
to the point that love is what you prefer,
and patience
is what you are willing to spend -
to get it,
to give it,
to be it.
It will be okay,
because love is what everybody wants;
they just need time enough to know it.

The Constellation of Leo

Gavin White's *Babylonian Star-lore, An Illustrated Guide to Star-lore and Constellations of Ancient Babylonia* tells us that the constellation of Leo (the Lion) "represents the ferocious heat of summer: its radiant mane stands for the overbearing radiance of the summer sun... The bright star at its breast" (our Regulus) "was known as the king star," here standing in for "the favorite of the goddess, to whom victory was always granted."

At one time while immersed in surfing the web's ever-changing waves, I learned that there was a time within certain cultural interpretations when this same star was seen as a bird (it may have been a dove) such that the constellation was associated with both: a lion and a bird. Only days after seeing it, I wanted to look at it again, but alas! It had been swept away. No matter how hard I searched, I never was able to find it again.

Ninth Lunation

	...gather seeds of light.
●	
●	
●	
●)	

161

Dark Moon
In Stone

"I'm glad for our earlier discussion of Avalon, Cat. I don't know what you or I did, but somehow, one of us triggered something really fine..."

As I left the library where I had been writing via email to my friend, I watched the sun setting behind one of the hills there – at that moment, the hill looked and felt like Glastonbury Tor. It felt good, as though I had been greeted.

Two nights later I had a very special dream. I was traveling – with my husband, a sweet old lady, and a young boy and girl. The kids had brought all their belongings with them, certain that their foster parents would kick them out for going on this journey with us.

We arrived at the camping spot we'd had planned before our trip began. It was a circular clearing almost like a courtyard, paved in weathered stone. In fact it *was* a courtyard, belonging to an old stone building turning out to be a cottage in Wales – our elderly companion's home. There were lovely vines and flowers growing all along the walls. These walls and courtyard were her garden.

A black raven was following us around. It was hungry, and I was really sorry I didn't have anything to feed it. Running along the vines growing on the walls was a little creature like a hamster or a rat that ran right up to my hand in a bold and friendly fashion. It was fat and sleek. I drew my hand back and it drew back from me. We both sat there, looking tensely at one another. I asked the lady what it was, and relaxed when she told me it was her pet. So did the creature; very soon we were becoming friends.

At about that time, the elderly lady and I walked through the open door of the cottage. The moment we went inside, I realized this was actually an old stone church I've seen in dreams before, located (my dreams would have me believe) somewhere in Scotland.

	Waxing Week
Gaiaday	
Sunday	Stone Henge, a solar temple and observatory that predicted eclipses accurately, served as both sundial and calendar; "it was begun 200 years before the Fall of Troy and finished in 1600 BC." (based on information in Gerald S. Hawkins' *Stone Henge Decoded*: p.140, bottom paragraph – p. 147; quoted portion taken from same source, bottom of p. 39)
Moonday	

The pews were of dark wood and smelled of lemon oil, in a long narrow building all in dark wood too. There were many large windows of leaded glass. The place was filled with sunlight and an air of peace.

She picked up what looked like a hymnal and we began looking at it together, but instead of song sheets it was filled with photographs of large stones, into which extremely simple designs had been etched. Each design was framed by a circular border, like scenes viewed through a round (rather than a rectangular or square) window. The first was just a circle with nothing in it. Another had a circle with a crescent moon at bottom, the points of the crescent facing up, in a third there were stairs, while in still another, a grouping of stars, and there were others showing details that I've lost.

I suddenly began to remember (while dreaming, though I couldn't after waking) places I'd seen or heard about while travelling in the U.K. By reading the captions under each plate, I was recalling them very slowly, one by one. These captions had a rhythm all their own. I was having trouble pronouncing them: not getting the rhythm had me missing the meaning itself, though as I kept trying the right sounds began falling into place. They had to be sung: this was, I saw over time, a song of welcome.

We turned the last page. I saw no photograph of weathered stone, only a greeting card in royal blue, with star-shaped cutouts showing silver paper behind, and one or two more circular scenes. In one there was a mountain like the one I lived on, in dark silhouette under a sprinkling of stars. Turning to the woman holding the book with me, I saw she had grown young. And I *knew* her – a cherished friend, black haired, with blue-gray eyes: my long-lost twin from a time shared long ago. Seeing her again filled me with joy, well beyond the wakening.

	Waxing Week
Mercuray	
Vensday	
Tiwsday	In Romano-Celtic settings, Mars would be invoked as a healer, his ability to overpower the enemy in battle being transferred to the sick person's struggle against illness.

Harvest Moon
A Glastonbury Adventure

While near Glastonbury some years ago, I walked right through something, all unaware. Notes taken read:

'8/25/03
An angry day. Abergavenney's Black Sheep had raised their prices two pounds above the advertised price per night, then let me stay for four before telling me at checkout: eight extra pounds gone, and for no reason. Then on the train tried to buy the same excursion ticket I'd bought in West Wales and was told yes, it's still available but no, it could not be bought on the train, only at the station; this while already underway. So, an extra seven pounds above and beyond: fifteen all told. Again, very angry – mostly at Black Sheep! Fairly uneventful trip to Street. Got there early and dropped pack off at hostel. Plenty of day left, so, still furious about money, decided to walk to Glastonbury (big mistake).'

Some have seen something hidden in the landscape around Glastonbury. How long has it lain there? No one knows! Throughout the Somerset area, a zodiac of sorts formed of giant earthen figures miles long sometimes utilize the river for their delineations. Twelve signs, together with two "guardian dogs" and a griffin form a circle 10 miles across and 30 miles around.

"The Great Dog of Langport lies along and south of the road from Street to Taunton. The Polden Hound is north of the Great Dog, along the Bridgewater road..." (Robert Coon)

'Followed a "public footpath" for a little while, hoping it was a short-cut, but then it disappeared. Tried to forge my own short-cut through a cornfield – got trapped behind hedgerows and panicked, then found a way out by water (the river Brue) and saw, far distant, the Tor: my landmark through the shrubbery.

166

	Waxing Week
Jupitay	Jupiter is twice as big as all the other planets combined; eleven Earths could fit across its equator.
Saturday	
Starday	

Harvest Moon

Very tired by the time I got to Glastonbury. Came out by the Roman Way, but didn't know it. Followed it along Wirral Hill.'

"**Pisces:** Two fish, tied together with a bond of love, between Glastonbury and Street. Leviathan swims along the River Brue lowlands, threatening to break the bond... Wearyall Hill forms the northern fish. The southern fish is on the northern edge of Street... Wearyall Hill, the summit..." is key. "...The Celtic stone, dedicated to St. Bridget... stabilises the dangerous tail fin of Leviathan, on the east bank of the Brue..." (Robert Coon)

'Reached the Folk Museum minutes after they had closed and asked someone near me for the way to Joseph's thorn tree. It was close by: "Wirral Hill" is "Weary-all Hill." A woman was already there, waiting for a large group to move on, so I sat and waited too. The tree was filled with ribbons (momentos of many prayers). While I sat, I had a lovely sense of communion with someone dear to me who had passed. Giving in, after some time, to hunger, I invited my departed friend to walk with me as I headed down the hill, to where I could see there was a supermarket. Got there ten minutes after it had closed. Grabbed an expensive piece of fried chicken, gave in to exhaustion and paid for a bus trip back to Street. Wrong bus: dropped me off far from the hostel while it headed up to Walton. Very angry, again with Black Sheep!'

"**Aries:** A lamb, sitting on the ground, its head tilted back looking west. This figure contains the town of Street. Walton Hill, south of Street..." is key. (Robert Coon)

	Bright Week
Gaiaday	
Sunday	
Moonday	"Monday's child is fair of face..."

Harvest Moon

'8/27/03
*Stopped to visit an encampment on my way out. Been meaning
to visit them all along (lovely fiddle tunes floating in from that
direction, two nights running). Several families, with two horses
and some wagons – no car – very curious about them. Between
jobs, but one about to start. Modern gypsies, not Rom; English,
living creatively in a pleasant way that coped handily with a harsh
situation. Very friendly horse (met him the day before, tied
beside the trail). They knew all sorts of sacred places in
Glastonbury, including some of the more hidden ones, and tried
to tell me how to find them. Really regretted I was leaving that
day. Funds said I had to. I still regret going.'*

"The initiatory pathway around the Zodiac begins with the two
Guardian dogs and then enters the Zodiac at Wearyall Hill in
Pisces. There are three...phases of the Quest. The first...moves
from Pisces..." "...the Search for Excalibur. The first duty is to
find Excalibur, carry it for a while, and then return it to King
Arthur. It is eventually placed in his empty hand at the
knighthood on the River Brue...Arthur uses the returned Sword
to alchemically transmute Leviathan..." (Robert Coon)

If anger was Leviathan, I wasn't great at transmuting it right then,
and a deep enough connection would not have cared about
missed flights or anything else, but something beautiful was afoot
then and there, all the same. In time, I completely forgot I had
been angry. All that I could recall was that sweet, sweet sense of
communion with my departed friend.

Rosamund

*Let there be beauty, and let beauty be upright.
The Rose of the World goes calling, calling.*

	Bright Week
Mercuray	
Vensday	
Tiwsday	"...Tuesday's child is full of grace..."

Harvest Moon

On the wind, the leaves are falling,
f *Falling like tears on the face of the Rose.*
She changes everything she touches,
and everything she touches changes.
Let there be beauty, and let beauty be upright.
Let there be love, and let love be light.
Overhead, a tune is weaving, weaving.
Deep within, the earth is singing,
singing like peace in the heart of the Rose.
She changes everything she touches,
and everything she touches changes.
Let there be love, and let love be light.
Magnify light in the beauty of night!
Call forth the beauty that blows
from the sorrowing love in the heart of the Rose!
She changes everything she touches,
and everything she touches changes.

The Word 'Friday'
That's 'Frigg' and 'Freya', *not* "friggin' Freya!"

'Friday' comes from a Latin phrase meaning "Day of Venus" (the Roman overseers' best attempt at translating the Anglo-Saxon 'frije-dagaz'). Despite much scholarly debate, no one can be certain as things are now whether the day was given over to Freya or to Frigg. While some say with great passion that the two are one, others demur, with equal conviction. This duo frustrated me so much I sometimes *did* feel like swearing. Every article I found on either of them seemed to contradict every other article I found, until, finally, I found in one a mishmash of many: none complete, all run together, no punctuation, sometimes cutting off

	Bright Week
Jupitay	
Saturday	
Starday	In second century Britain, a soldier's inscription attests to Ceres' role at the time. She was "the Syrian Goddess," the universal heavenly mother, Virgo, the virgin mother of the gods: inventor of peace, virtue, and justice, weighing "Life and Right" in her scale. 173

in mid-sentence. Anything online having anything at all to do with Frigg or Freya, from ancient figures to game characters to modern puppies had been collected here, it seemed. Had some frustrated person out there taken this down and dirty approach to figuring out who the heck these two are? I seized on it with gleeful gratitude to the lady whose (forgotten) name was attached. This became my source, where I learned:

Frey(j)a was born to a sea god (Njord), and an unnamed female deity from among the Vanir tribe. Her name may be drawn from Slavic or Old Norse and mean "Lady"/"The Lady," or "High and Courteous." In the poetic gathas, the word fryá means "an intense feeling of passion or love," so she may be a personification of the same (fryá/Freya: "Dearly-Loved"). She was goddess of everything: love, beauty, sex, and fertility; the harvest, war, death, and birth; wealth, gold, knowledge, divination, and other magics. She was queen of those "female battlefield choosers of the slain" we know as Valkyries, and would ride a broom to sweep away clouds whenever they piled up.

Frigg(a)'s name is Norse and it too means "Beloved." She was an Aesir, and part of an older, "more elusive" tradition than Freya. As queen of the goddesses, she usually had a group of women around her, and was often called on: for safety in child birth, the protection of children and the family, and preservation of family bonds in times of strife. Seen as the Mother of all, she was called on too for peace in other tribes, clans, non-profits or small companies.

Lovers would send prayers to her. Ull sees her as the Norse goddess of marriage, love, destiny, chastity, fertility, motherhood, and wisdom, as well as "what the Norse considered 'women's work,' such as house-hold management, spinning, and weaving." She knew everybody's destiny, but would never reveal it. The

	Waning Week
Gaiaday	20 percent of the earth's atmosphere is formed by plants when they produce oxygen.
Sunday	
Moonday	

175

stars that form Orion's Belt were called Frigg's Distaff at one time. Upon these, she wound the threads of fate (and wove the clouds that brought the rain, greatly impacting the fertility of crops). She spins the sacred distaff of life, while personifying the sky, clouds and earth.

Norse myth gives us the feeling that Frigg was the protector of the sanctioned marriage while Freya protected unsanctioned unions. Freya was worshipped by lovers, but when marriage was solemnized it was great Frigg who was invoked. Both goddesses were married to Odin, and had a brother named Freyr. Both coveted gold, jewelry and the love of men; both had numerous lovers. When Odin was exiled from Asgard, Frigg carried on an affair with his brothers (Vili and Ve). In addition to presiding over the realm, they regularly slept with her until Odin returned. At that point, she returned to his side, and was his wife again. Both women were Viking Age practitioners (Volvas) of a form of Norse magic called Seidr (discerning the course of fate and calling new events into being while working within its structure, through the use of symbols). Both wore cloaks made of bird's feathers (falcon for Freya, raven for Frigg) that had magical powers, allowing them to fly across constellations and between worlds in bird form. Ravens, hawks and falcons were sacred to both goddesses.

"Frigg or Freyja, in all cultures she symbolically represents the perfection of female power: wife of Wotan or Odin, driving her chariot pulled by cats..." and owning "...a boar, named Hildisvini." Some Germanic and Norse peoples recognized only the one and some, only the other; while in some stories, the two are merged. I find it interesting that in an earlier time, the Norse and the Germanic peoples were but one collection of Teutonic tribes. Frigg and Freya give every appearance of having evolved from the earlier Germanic goddess Frija.

	Waning Week
Mercuray	"...Wednesday's child is full of woe..." I've no idea why that might be, but do you suppose it has anything to do with Woden?
Vensday	
Tiwsday	

177

Harvest Moon

"...By the late Viking Age (when our sources were recorded) Frigg and Freya are only nominally distinct figures..." (who may not have completed whatever separation process they seem to have been undergoing). Whether 'Friday' comes from Frigg or Freya never did become clear. Perhaps 'Frije' is neither? What do we *really* know about the earliest Anglish? I recently rediscovered a version of the original mishmash in my files (minus all puppy and most game-figure parts) entitled "Frigg/Freya Ocean," almost seventeen pages long. It included the web page address. When I tried to go there, I pulled up a job/worker matching website based in India that said "page no longer exists."

The Constellations of Virgo, Libra, and Scorpio

Given the connections between lions and "The Lady," it's interesting that Virgo follows right behind Leo in the sky. The word is Latin; it means self-contained or self-sufficient, so care from this "virgin" would simply be an expression of kindness or creativity and would not come from any need for others (5). Virgo as an icon shows us a self-sustaining nurturer: mother to all, gliding through the sky as a woman holding an ear of grain. It was formed from two Babylonian constellations: Mul Apin (The Furrow) with bright Spica (Latin star name meaning Ear of Grain) in the eastern part; and to the west Ab Sin (the Frond of Erua), associated with fertility and depicted as a woman holding a palm-frond (3). We sometimes still see this motif, even in late depictions of Virgo, when in the Middle Ages, she would occasionally be associated with the Blessed Virgin Mary(3).

Many of the goddesses linked to Virgo (Babylonion Ishtar, Egyptian Isis, Roman Ceres, Greek Demeter, etc) were ladies of the harvest. Stories from Greece and Babylon tell of someone

	Waning Week
Jupitay	"Thursday's child has far to go..."
Saturday	
Starday	

dear to a harvest goddess being trapped in the Underworld, but then (in one, at least) allowed to return for six months out of the year. Interestingly, Virgo is only visible in the skies from March through August – six months out of the year (5). Among those linked to Virgo we find goddesses of justice also (Greek Dike and Astraea, Roman Erigone). Some stories say the scales of justice (gliding by her side as Libra) are hers and that she is "literally holding them for eternity in the stars" (1).

In early Babylon, "Libra was Mul Zibanu (the Scales), sacred to the sun god Shamash, patron of truth and justice." "The symbol for it: ♎ , might well have come from the hieroglyph Akhet: 'the place of sunrise'"(2). Libra has always been associated with fairness, then and now. "During the sun's transit of this constellation the days and nights are equal and fair...Libra balances night and day" (2).

 "Just to the east (left) of Libra of a June evening, the most beautiful constellation can be seen coming into view low in the southeastern sky: Scorpius, the Scorpion. Most scorpions have two large claws in front" (4), but not this one. In fact, it had its claws taken away! In Arabic culture and in very early Greece Libra was not seen as the Scales but as "Scorpio's[missing] claws, hung upside down" (2); but "a few thousand years ago, the Romans decided that there should be 12 constellations in the sky instead of 11" (4), so they took the two claws from Scorpius and turned them into Libra's arms.
 "Two of Libra's stars still bear their Arabic names: Zubeneschamali and Zubenelgenubi ('Northern Claw' and 'Southern Claw'). Zubeneschamali has a faint greenish tinge: the only green star one may view without telescopic aid" (4).

For quoted sources (numbered here) see my Back Pages, under Bookish, 9th Lunation.

Tenth Lunation

Her mystical form...

181

Dark Moon
The Chairmaker's Lass

Having this dream was like watching a movie. The setting for it was in a town by the sea. In the beginning, there are lots of people down on the shore, swimming, sunning and playing, including a little girl and her parents. A certain musical theme becomes associated with this girl, reappearing often, as such themes in movies typically do. We hear it play whenever she is about to lose something. In this dream, that happens a lot.

This time, she is playing with her kite, but the wind takes it from her and carries it up over the cliffs. She runs after it, but never can get it back. While off chasing it, a rogue wave crashes over the beach and washes almost all the people there out to sea. Its fearful power will color her whole life, for she has lost both kite and parents in the same few breaths.

We see her later as a child at school, with all her classmates around her. One of the boys is a special friend; he is quite the artist. He has woven a really charming (miniature) chair out of sticks, which he is so proud of he puts a plaque on it to show it was made by him. With a flourish, he gives her this chair. We see him going through the years, always making chairs, big ones people can sit in, and he keeps putting special plaques on them just like he did with that first one. The plaques, too, are beautifully decorated. His chairs become very valuable – people try to imitate them, but theirs are never quite as good.

We see another pair of kids at the school who are both always getting involved with higher-up types (running errands for the teacher, etc). We see them moving through the years. They keep running across each other at unexpected times – each still occupying a niche somewhere within officialdom. Always in competition with one another, their antics serve up many laughable moments. They must have been written in as comic relief.

	Waxing Week
Gaiaday	The earth is round because gravity pulls matter into a ball, but carries a "spare tire" at its waistline on account of its rapid spin, flattening the poles and bulging out the middle.
Sunday	
Moonday	183

Apple Moon

Later there is a scene in which the orphaned girl (now a lovely young woman) and her dear friend (the maker of chairs) have become lovers. He has made her another one. Unlike the rest, the plaque it bears is turned, and hidden against the wood – only its plain back can be seen. It bears lovely decorations, even so, but they are only for her, so they must face inwards. They walk together on the beach as they plan their wedding. Just then a big wave comes right to them, and leaves what looks like a ring on the sand at their feet. Picking it up they see it's not one ring – it's actually two. They gaze at these in wonder and begin slipping them on. Out of the blue an enormous wave comes and takes just him, leaving her alone again.

Large waves come often in this dream, taking people away. And there are intermittent flashes of another place altogether, with kids playing on bikes and crowds of people meandering through the streets. It's a completely different town, yet seems related in some way.

As the scene changes we see the orphaned girl again, quite elderly by now. She is seated in the same chair her beloved chair-maker gave her – the last he ever made; surrounded by grandchildren and even a great-grandchild or two. "Yes, I still have my chair," she says. "Don't think it has no decorations, just because you can't see them. I can feel them; this chair has the loveliest decorations of them all."

The dream finishes up with a final scene: we're back in that other town. There are kids racing along on bikes and people walking by– maybe stopping to chat or look into a shop window. Suddenly and with one accord, everyone in every part of town stops whatever it was that they were doing. Lifting arms over heads they begin to dance, slowly and gently swaying in place. Inexorably, all is transformed. The winding streets and

	Waxing Week
Mercuray	
Vensday	"...Friday's child is loving and giving..."
Tiwsday	

185

shop windows become undersea cliffs, the dancing people are seaweed growing tall in offshore kelp beds, the playing kids are darting fish. Nestled together on the sand below, we see two starfish, one arm of each intertwined with the other, and at the tips of those two arms, the two rings that came to the chair-maker and his lass on the beach that fateful day.

The Mermaid Chair

Did you know there really is a special chair? It's over 600 years old now and has carvings of fish on the seat. It was formed from two medieval bench or pew ends. Worked into the whole there is a depiction of a mermaid, comb in hand, admiring herself in a hand-held mirror. When people speak of it they call it "The Mermaid Chair." You can see it for yourself if you are travelling through Cornwall. It's in a church in the village of Zennor, about five miles southwest of St. Ives. If you go out into the churchyard on the south side of the tower and look up, you will see a clock bearing an inscription dated 1737, and on the bronze clock dial, again, the figure of a mermaid.

No one really knows why these carvings are here, though there are a number of stories connected with them. They may have been concocted to explain a few things, the truths of which lie forgotten – or...? Some have links with the patron saint for whom the little town is named and some do not, but if taken as a group patch-worked through time you can see inside them the evolution of a soul... Or is it a culture? It's just as likely to be both.

Look in my Back Pages (pp. 268-71) for the whole collection woven together as one. It was in the Author's Note at the back of Sue Monk Kidd's book *The Mermaid Chair* that I first heard about chair, saint, and princess.

	Waxing Week
Jupitay	
Saturday	
Starday	Pluto is the higher octave of Mars, destroying in order to renew. It brings buried but intense drives to the surface and expresses them. Governing great wealth and power, it relates to any enterprise that involves diving or digging under the surface to bring truth and health to light.

Apple Moon
Pictures from the Mermaid in My Mind

A poem rose out of me once, quite suddenly, as if it were a memory. I have always cherished the glimpse it gave me of itself. The speaker appears to have been a mermaid. I think it fits very nicely here:

A golden beach,
starfished...
 no, I mean sand dollars – dried out, bleached white,
like those little clam shells;
golden mica, and the setting sun –
 that orange golden light is everywhere: *a golden bath.*
Feel so good, so tired...
stretched out, face down, hair flowing everywhere
...down into the sand
like little roots.
So good,
so tired;
don't move.
So tired so tired so tired...
Oh yeah, a tail...!
Stay in touch with the water.
Hold to that rock
and stay down with the water.
Oh, it's leaving.
Hey, can you carry me?
Down to the water.
I'll give you three kisses and a song –
kisses,
so you'll always remember me;
a song,
so I'll always remember you.

	Bright Week
Gaiaday	The earth's rotation is gradually slowing, by about seventeen milli-seconds every 100 years.
Sunday	
Moonday	

Apple Moon

Sometimes I'll sing the song
and maybe you'll hear the echoes.
They'll help you to remember what's good.
When you hear them
you'll stand up straighter,
and you'll try harder.

I wrote one myself many years earlier, which would also go
nicely here:

Avalir

Island of the birds,
gull's cries through sea-spray misted;
island of mermaids,
and sea-serpents, long tails twisted.
Purple shadows fill the turquoise canyons
and clear waters sway the little green strands,
where sea-fish roam in dark sea caverns
in hidden underwater lands.
In the thick white mists on Avalir,
where abundant beauty thrives,
majestic oaks crown island sheer
and bless a thousand lives.

	Bright Week
Mercuray	
Vensday	Both Greece and Rome believed Venus was two stars: the "Morning Star" (Phosphorus or Lucifer), and the "Evening Star" (Hesperus or Vesper).
Tiwsday	

Apple Moon
Halloween

The word for this holiday comes from Catholic "All Hallows Eve," but a much older celebration (which seems to be enjoying a resurgence again) was once held in Celtic lands on the same day. It was called Samhain (SAwin), a Gaelic phrase meaning "Summer's End," and an abundance of traditions were enjoyed at this time of year. There were bonfires, plenty of food, and people playing all the games of summer one more time, supposedly so as to console lonely ghosts, who were believed to walk among the living on this day (and one other: May Eve, called "Beltane") only. Efforts were made to keep these ghosts at peace, for they were only old loved ones who had passed on. This was the perfect time to ask them questions (usually about life or love).

In medieval England lots of lanterns would be hung, carved out of vegetables. Turnips were a popular choice, but never pumpkins, for Europe had none. Candles, too, were lit a lot on this night, for lights were needed to guide lost souls, cheer up estranged ones, and scare off evil ones. Children would wear masks and go from door to door, "souling" which means they sang songs to beg for special treats (called "soul cakes"). These were not for themselves but for lonely ghosts, and if none were forthcoming, the children would play tricks and then say a ghost had done it.

There would be revelry indoors, too in lordly halls, with games and a mummers' play. Towards the end, seven guests in masks with baskets would walk through the hall, chanting "Souling, for soul cakes we go: one for Peter, two for Paul, three for Him who made us all. If you haven't got a cake, an apple will do; if you haven't got an apple, then give a pear or two; if you haven't got a pear, then God bless you." Chanting in this way, they'd come up to each table and be given a fat soul cake (really a cross between a small cake and a cookie) from a special platter,

	Bright Week
Jupitay	
Saturday	"...Saturday's child works hard for its living..."
Starday	

193

Apple Moon

or pieces of fruit that are decorating the table, unless someone wanted to stir up some fun by looking reluctant. This was worth doing, for the soulers could then threaten that person with amusing punishments. The filled baskets would be left sitting around the bonfire and later donated to the poor.

Divination games were popular at Summer's End, too, especially those concerning love. Bobbing for apples was one. First, you name an apple for your crush. Capture it at first bite and you'll be happy together for years to come, second try: you'll be madly in love, but only for a little while, third: she'll think you're a pest and hate the sight of you, four or more: run fast, run far – do *not* look back. And there were apple parings. Using a whole apple and a small knife, cut a long spiral ribbon of apple skin. Throw it over the left shoulder and watch how it lands. It should look like the first letter in a love-to-be's name. Crowdie was another fun one. When six revelers around a bowl of apple sauce mixed with sweet cream containing two each of clean marbles, coins, and rings dig in, what will they spoon up? A ring (soon to marry), a coin (ever wealthy), a marble (cold and single), or nothing (sweetly uncertain).

(Festival material is drawn from Madeleine Pelner Cosman's book: *Medieval Holidays and Festivals*, pp. 81-86)

Soul Cakes

Recently, I found a recipe for Soul Cakes. It never said how old it is or isn't. When baked, they really did seem part cake and part cookie, and they'd a delicious though also an unusual flavor. They weren't all that complicated to make. You just

194

	Waning Week
Gaiaday	
Sunday	"...and the child that was born on the Sabbath day is blithe and bonnie, good and gay..." In Spanish, Sabado means both Saturday and Sabbath, but in the English speaking world Sabado (the Sabbath) became Domingo (Sunday) when Rome took our day of rest away from Saturn and gave it to the Sun.
Moonday	

195

Apple Moon
Soul Cakes

Take:

¾ cup caster (baker's) sugar
1 ½ sticks of butter   *beat with sugar until mix turns white*

2 eggs   *lightly beat eggs, adding a little at a time, always beating*

3 ½ plus cups self-rising flour:
(3 ½ cups flour + 1 Tb baking powder + ¾ tsp salt)
½ Tb of ginger
2 tsp of cinnamon
¾ tsp nutmeg   *mix the flour and the spices together,*
½ tsp allspice   *then add to previous mix*

1/3 cup raisins   *fold in raisins*

2 Tb of milk   *heat milk and gradually blend into dough*

1. On a floured surface spread the dough out ¼ inch thick. Cut
   or form it into 2 ¾ in rounds and place them on a baking
   sheet or plate covered with baking paper.

2. Prick them lightly with a fork and make a deep cross on each
   with the back of a knife. This symbol represents a soul newly
   freed from Purgatory.

3. Preheat oven to 350 degrees and bake for 15 minutes.
   Remove from oven, sprinkle with sugar and let cool on rack.

	Waning Week
Mercuray	
Vensday	
Tiwsday	Mars became the god of war and also an agricultural guardian, representing military power as a way to secure peace. He was second in importance only to Jupiter.

197

Apple Moon
Wednesday/Wodendagaz... Who is Woden? Not Odin!

Both deities may have come from the same proto-Germanic source, but the Teutonic tribes to which each once belonged saw many changes in their languages, beliefs, and tribal identities, especially during the time of the long migrations (400 – 800 CE). Those who remained in the north we know as Nordic, and those who journeyed farthest we call Germanic, but most of us have forgotten now that at one time they lived side by side, and spoke the same language.

Woden may be the best known of the high gods belonging to the Anglo Saxon tribes. He is identified as both progenitor of their royal houses and builder of the great earthworks in England. Like Odin, he was considered king and Allfather of the gods: creator of runes, esoteric and magical; but not quite like Odin was it said that " *Woden id est furor* " (Woden is fury and inspiration; Woden is madness incarnate). One could become "Wod-mad," working feverishly at a task until it was either completed or the enraptured one collapsed.

Less refined, more dangerous, coarser was Woden; more primal, more hungry: ecstatic mind-madness, fully embraced. He knew well how to navigate such fury. With his wild black horse and his carrion crow, he belonged to the fifth and sixth century meadhall culture of death, war, and the raven that feasts on the hanging corpses. Human sacrifice was the norm, before battle to gain favor and after, to give thanks. Coldly capricious, it was Woden's task to select which of his own men must die. The chosen one was trampled by a horse while being speared by its rider. The Romans equated Woden with Mercurius, the Trickster – a much milder being – and gave him Wednesday to oversee.

	Waning Week
Jupitay	
Saturday	"Jovial" once described someone born under Jupiter (jolly, optimistic, and buoyant).
Starday	

Apple Moon
When Astrological Aspects Overly crowd or are Absent

Too many of these in a chart and the chart is describing:

- *conjunctions* - a focused person, motivated from within;
- *sextiles* - a communicative, creative, and expressive person;
- *squares* - an action and crisis-oriented person, motivated, willing to tackle obstacles but not blindly accept things, energetic, and having both strength of character and will;
- *trines* - an easygoing person, comfortable but talented;
- *oppositions* - an insightful but contradictory person looking for deeper meaning, who usually seeks growth and balance in life through relationships; though unfocused and inclined to project his or her own qualities onto others, more able to be objective through same.

Too few of these in a chart and the chart is describing:

- *conjunctions* - a scattered and flexible person;
- *sextiles* - someone who avoids creative pursuits, is less self-expressive, and isn't very communicative;
- *squares* - a person who lacks motivation, takes the easy way out, avoids confrontations, and is rather habitual;
- *trines* - an unbalanced soul, conflict-oriented and driven;
- *oppositions* - a subjective person who doesn't need a partner to mirror his or herself, self-contented and not as concerned about growing through relationships or broadening his or her own perspective; though self-reliant, s/he has a hard time seeing the other side and s/he easily loses perspective.

I found all this at cafeastrology.com. You can find an in-depth look at aspects in general in my Back Pages (pp. 271-4).

Eleventh Lunation

	From unborn to born...
●	
●)	
	201

Dark Moon
Making a Movie

As you probably have noticed, a dream or a meditation usually leads each lunation. I arrange them that way because they're meant to reflect dark moon time, and it was to me that almost all of them came. But there is one exception. I kept quiet about this while I was sharing it not wanting to spoil the effect, for now is the better time to let you know: the dream that was so like a movie actually came to Bob and not to me. Yet well before he had his Chair-maker dream, I had *this* one. I never talked to anyone about my dream until after he'd had his.

In the dream I had, we were all hard at work on some big project. I think we were making a movie. It might have been *the Chairmaker's Lass*, though at this stage we haven't got any specific name for it. There is a boat (or maybe it's a very large rock), and waves are being simulated. Big ones keep pouring over the set, coming at us again and again (the director is looking for the perfect shot). Will they wash us away? They're supposed to go many feet over our heads, but we're supposed to cling to the rock or boat or whatever it is (just a simulated bit of floor – it probably doubles as both) and still be there when the wave is gone.

Some mutual friends have joined me while we work. It's summer, the air is hot, and they're quite tanned; several of them are dressed in white shorts and light summery tops. There was something I was supposed to go shop for and when I didn't come back with it they went looking for me, but I wasn't where they expected me to be. So they came back to the project and found me still working there. I had gotten so involved in building the set that I'd forgotten about shopping for that thing, or else I remembered we had built it into the set already but by accident had let it get hidden somewhere underneath some stick-on stuff. We started searching for it under the guise of doing more set work, and were peeling off paper from some place where it had been stuck on. It was like peeling a label off, except it came up not sticky. It was rainbowy,

	Waxing Week
Gaiaday	
Sunday	
Moonday	In 1918, astrologer Sepharial proposed the existence of Lillith, Earth's "dark moon," though the same name is also used to refer to the axis of the actual Moon's orbit.

lasery stuff, mostly grayish (muted colors), with very little shine: used on the set to seal in water.

We were working intently when out of the blue some man snatched up the piece I'd been unpeeling and crumpled it into a ball before throwing it back. I was so angry at first I was almost yelling at him, emphasizing certain points by shaking my pointing finger at his face and even tapping him on the chest (!) He took it amiably enough, spoke apologetically, said it wasn't me he was after but the bank this project was to benefit. He was furious with them; by breaking a certain promise they had hurt him badly.

I could understand that. I wasn't mad at him anymore. But I did want him to understand the fruitlessness of his behavior. We had gone our separate ways by then. I found him untangling strands of something meant to be sea kelp, and took a place next to him. He sounded Peruvian, and was darkly Indian. I said,

"You know, the only person you hurt by doing that was me; this project will never need what I was gathering up."

"Oh dear," said he, "am I never to live down my misdeed?"
I relent, and say,

"I don't really think you hurt me all that much. It was more the idea that I objected to."

I meant to add, "If you're so upset with the bank, why not go after *them*?" But by now the drumming is starting up; the beat is too intoxicating – no one can stay away. We all leave our previous conversations or occupations, drifting off to dance.

Honda Shadow Spirit

How do I tell of Honda Shadow Spirit? It's too hard. None of the words I know convey his wackiness, his warmth, his relaxed and open way of embracing life. He was a half-grown black and white tuxedo cat, who appeared at our house one cool December

	Waxing Week
Mercuray	Did you know that Mercury is wrinkled? It has wrinkles all over its surface.
Vensday	
Tiwsday	205

day. *I* couldn't adopt him; I didn't want to make my own cats jealous.

I started carrying him away; he just lay in my arms, purring. I set him down atop a park fence, and we looked across the park at all the houses on the other side. "Now where are your people? They must be amazing – love is all you seem to know. Shouldn't you be going home to them?" Then I walked away.

He was right behind me. No one we met along the way would have him, so when we got home I picked him up again and carried him (totally relaxed, *still* purring) inside to James. "How about you, do you want a cat?" Another no, but a few nights later, the little guy was still with us. He had moved to the front porch and James saw him sleeping on the couch there, looking cold. So he took the kitty in and named him for his motorcycle: "Spirit" for short.

Our new cat acted like he already knew every corner of the place. His fearless friendliness, curious green eyes, and big ears set low and wide reminded me so much of a character named Merlin (in that Hulu series of the same name), I sometimes caught myself calling him that. He was supposed to have a relationship with James only and not with me. What was I *thinking*? My own cats (who never, ever consent to come inside) didn't have a chance. But they couldn't resist him either. Soon, he was everyone's friend.

Of course, he loved to play. He specialized in doing the unexpected (like plopping down on our feet when we weren't paying attention). I was walking down our hallway one day, absorbed in thought and looking down at the ground, when my eyes were caught by a blur of movement. It was Spirit, directly in my path on his hind legs, front legs and paws spread wide (then on all fours again in the blur of a second). And there was that other time when I was coming home from work and parked the car, with a pile of mail in my lap. I sat there reading it for a little

	Waxing Week
Jupitay	
Saturday	
Starday	Stars have very long lives. Those the size of our Sun will have been several colors before all is complete: yellow, red, white, and black.

while, until without thinking about it I found myself looking up – into Spirit's green-gold eyes. He was crouched as close as he could get against the windshield, staring at me! I rolled down the window and he strolled right in - only for a second or two. For such a solid, muscular guy he had the softest fur. Even though he loved to snuggle, he was a one-lap cat – James's for a long while, and later mine. We would both end up fast asleep somehow, every time.

The year went by quickly – it was December again. On the first day of winter, we went for a drive. Looking back, I wish we'd taken Spirit along. He probably would have liked it. We'd had to lock up the house with him outside – there were bugs to be rid of and poisons to set off. We headed west, and had a long, beautiful drive through unexpected rain. When we got home, we saw it had been raining there too.

Everyone except Spirit showed up to get their rather late dinners. I was sad not to see him, but there were other times when he'd taken off for a few days. I expected him back soon. James didn't see it that way. I didn't know he'd had a mental view of our cat lying on his side somewhere, all mangled up. I didn't agree when he said he thought Spirit was dead. I kept watching for him, as days turned into weeks. Then something extraordinary started happening. Powerful waves of feeling would descend, just out of the blue. In my mind I would see his earnest little cat face, almost feel his soft warmth in my lap, and with it such deep love. I sank into that love and sent those waves right back, staying very still and soaking in them. I hoped I was sending him healing energy that could help him if he was hurt or sick somewhere. I sent him all the love I could.

A few nights later I had a dream. I was in the backyard of a brownish-green house with tall cement steps leading to a white back door, and I was playing with Spirit! The hair was patchy or

	Bright Week
Gaiaday	
Sunday	
Moonday	Both Hecate and Diana Trivia ('trivia' means three-way) were goddesses of crossroads. Choose carefully – the wrong road may lead into sorrow, while the correct one finds you joy or grace.

gone from three quarters of his left side – only scar tissue there – but we were playing. He seemed happy, and full of energy. It was *so* good to see him. I hoped he'd come home soon, now that he was healed. Later I reached out in my imagination to visit some more, and found him deep in play – with someone else! That's when I understood: he wouldn't be back. He loved those other people too! I was just happy to know he was in good shape somewhere, and having fun. It was over so soon, and so unexpectedly, but how beautiful it was – our Year of the Cat!

Dihunta

The sun rose, spilling out its beams,
pouring them across the land like golden water.
Is life only what it seems?
We're all settled here, in our ways, our times:
grab our iphones and get to work on time.
"Gotta run! I'll catch you later.
Gonna catch up on some schemes."
Won't you do your heart a favor?
*Tell me **now,** what makes it sing?*
Are there unicorns
in the sea? What voice,
the pale moonlight?
Just out beyond the edge of sight,
tugging at our heart's core,
unlooked for –
love beyond our wildest dreams.

	Bright Week
Mercuray	
Vensday	
Tiwsday	At least two animals have been named for Mars: Martius Lupus (a species of wolf) and Picus Martius (a woodpecker, species unknown).

211

Things Chief Joseph Said

Chief Joseph said so many things.
I know those things he said of us are true.
We love to buy and sell –
goods, dreams, even the earth itself;
if we could, the sun; if possible, the sky,
and adulation goes to those who can bid high.
Listen with me, and see through different eyes:
trees that give, and while they do it, care;
rocks that never forget; the fragrant earth,
cradling children that clothe her in color.
A healer is she, our mother, loving and feeding
all of us, for all our lives,
even when we
are at war with her...
At war? Ah, these are different eyes! And yet
to uproot and move on, to pave and abandon
is fashionable – a daily reality.
How many of us are close to family,
whether mystical or mundane?
My father is buried
far from his father,
and I had
no blessing from him,
no sense of home.
I hunger for the quiet places, wild, uncaged.
"Where are the thickets? The sweet, sparkling waters?
Where are the eagles? The fresh, clean air?"
I dream of the wild horses my great-grandparents knew.
And my great-grandchildren – will they dream of tigers?

Yes truly, life is precious, and at the same time so fragile. All of us are rich. I hope to spend mine well. Let me *live* until I die!

	Bright Week
Jupitay	Jupiter's deep oranges and browns are carried from the planet's interior to the surface on very strong currents of air. The darkest parts are the hottest.
Saturday	
Starday	

I do love this time of year. Earth finally cools down, and the air can play at being intoxicating. Autumn leaves are everywhere, looking gorgeous and crunching delightfully underfoot. My mother loves to crunch leaves too – it's one of my earliest fall memories. The loveliest fall leaves still remind me of her. And then there's Thanksgiving with all its partying and great food. The Navaho name for it is Keshmesh Yazhie, "the Little Christmas." A practice Christmas? I can see their point.

Lamb's Wool Cider

With thoughts on this day, I play with a recipe from the 1300's, a hot drink calling for apples and wine. I used sparkling cider and it was decadently sweet, but it tasted so good I boiled more the same way to seal and keep as apple sauce. The *real* taste needs wine, I suppose (I did that later, and it was lovely).   You just

Blend 2 quarts of apple juice with 2 quarts of white wine. Gently warm 3 quarts of the blend, setting 1 quart aside.

To the 1 quart add:
½ tsp ginger
¼ tsp cinnamon
8 tsp nutmeg
12 small apples (peeled and cored) that you first
broil to bursting, then boil, simmering until shapeless.

Blend 2 cups heavy whipping cream with
2 Tb brown sugar and
¼ tsp salt, and whip into peaks.

Serve the warmed cider in tankards, cups, or bowls, and add the hot apple blend to each with a dollop of whipped cream.

	Waning Week
Gaiaday	
Sunday	
Moonday	Luna is often depicted driving a two-yoke chariot drawn by horses or oxen, one black and one white, "for the moon travels on a twin course with the sun and is visible both by day and by night." Sometimes Hecate is the one who drives.

Hunters' Moon
Nerthus and Monday

Nerthus never left her name on any calendar of ours. Her worship was left behind in the mid 5th Century, while the Anglii were still a'wander. But there are writers who say she returned in male form, as Njord (Niord) the Vanir god of wind and sea. Or was she his never-named wife? One way or another, Frija was their child, just like Freyr (Frija's brother). Is Nerthus who Frija used to be?

She was a Teutonic goddess very long ago. Each year there was a festival where she would travel in a chariot pulled by two white heifers, escorted by the priest (so I say Nerthus belonged to the moon). No one was allowed to take up war or bear arms during the festivities. Even iron tools were locked up during the goddess' journey. It was good luck for those settlements she visited along the way. At the end of the festival, the priest would guide the chariot to a sacred lake, so that Nerthus might bathe. Her chariot would be covered with a cloth. After the selected slaves had bathed her in the lake, they were ritually drowned, for having served the goddess once, they must become forever hers.

There are those who say Freyr is somehow Frigg, thrice-burnt and thrice-reborn. The connection is not impossible: he was Frija's brother and Niord's (Nerthus's?) son. A regal visitor did grace my imagination one day. His ornate self-introduction could explain it all. (Greetings, Niord, son of Nerthus, husband of Frigg, father of Freya; hail and well met!) Ah, but what can the imagination tell us? It is all play!

So Nerthus is Anglish. Did she belong to the moon? If Frija had Friday, Nerthus should have had Monday – only she was too early – she had gone out of style.

Parallax

The apparent movement of one star against a background of

216

	Waning Week
Mercuray	
Vensday	Venus is associated with harmony, resilience, beauty, refinement, solidarity, affections, equality, sympathy and unity; also pleasure, comfort and ease, romantic relations, sex, marriage and business partnerships, the arts, fashion and social life.
Tiwsday	

more distant others has proven useful to astronomers. When measured, it can tell them how near or how far away the star is, through what they call stellar parallax.

Have you ever looked through one eye with the other closed, then closed that eye and looked through the other one? You stayed very still, but what you saw seemed to move – from side to side, as you switched from eye to eye. That's parallax too: a view of the same thing from two different locations (two eyes, a small distance apart).

Stellar parallax gets answers through geometry, and geometry relies on triangles, so how does this triangle in space get formed? By the earth on one side of its orbit, with data gathered at that time, the earth six months later (at the opposite side of its orbit) with comparable data gathered then, and lines of sight drawn from each of these points to the star itself. Using this triangle, two of its angles are measured with the one (earthly) side connecting them, using methods trigonometry initiates would understand.

Parallax also plays a role in producing three-dimensional images. The key is to capture two-dimensional images of a subject from two slightly different viewpoints that correspond closely to the spacing of the eyes, and present them in such a way that each eye sees only one image out of the two. The popular 19th Century stereoscope, the modern view-master toy, today's 3D televisions and even virtual reality gaming headsets all use these double images to attain their 3D effects.

There are many uses for 3D imaging in science and medicine. Scientists use 3D images to visualize molecules, viruses, crystals, thin film surfaces, nanostructures, and the like: so many things, not directly visible under the microscope because they are too small, or are embedded in opaque materials. A very fine online article (that apparently was not, after all, written by NASA) taught me all this.

Perhaps in these last three lunations, I've used parallax too: Avalon and Glastonbury, the seaside dream town with its

	Waning Week
Jupitay	
Saturday	
Starday	Neptune, the higher octave of Venus, is pale blue on account of the ammonia gas that swathes it. Uranus, the higher octave of Mercury, is the same color, for the same reason.

undersea twin, the same dream as movie and the one where a movie like it is produced, Spirit with us and Spirit greeting us from afar... If life were two-dimensional, what would it look like in 3D? What *is* the fourth dimension? Or do we know that already – is it motion or maybe time; should we be talking fifth dimension instead?

Something is here – I feel but don't see it, can sense but can't say it. Maybe it just won't fit into words. Who can stuff five dimensions into two? Okay. It's not meant to be said. Be very silent and very still, so there are no words, anywhere. When you have silence inside, then you can know. Not saying anything, you can know everything. Know it but don't say it, not even to yourself. That's when you might lose it. Sounds live and breathe, but words are two-dimensional: they keep us pinned, right here...

So in wordlessness, we can know what is otherwise unknowable. And have you ever noticed how sometimes we accomplish the most when we are at our most still? Ideas flow clearer, things connect better, tricky things work out at last. Hence the wisdom of not trying too hard: when we let go, become still, we can usually accomplish what we otherwise could not. Then that fifth dimension – is it silence, or stillness? Perhaps it is neither, but rather the ground within which both of these form.

Autumn

Autumn's leaves have turned to gold,
dancing, spiraling,
swirling down –
lost in the hazy-golden noon.
Such peace!
The calm of Autumn is on the land,
beautiful in the late-shining sun.

Twelfth Lunation

In sweet darkness lies...

Dark Moon
Le Cunamh De

I dreamed I had a challenge to solve, related to certain authority figures that wanted to interfere with me and were looking for a way to justify doing it. A phrase was ringing through my mind as I awoke: "Le Cunamh De" (Irish Gaelic, pronounced le koonev jay; it means "with God's help").

In the dream I had moved away from the land of my birth to a neighboring country, in response to an invitation from a group of people living there. They admired the skills of the people to whom I belonged and had requested that one of us come and teach them these things. Any one of us could have helped them, but I was the one with the fewest involvements at home so it was I who went. Of course I returned home often. My people were dear to me – dark-skinned, highly creative, very loving, and of quite a different culture: gypsies, perhaps. My frequent crossings of the border made the guardians there nervous.

"Oh, please let me come and go. I won't bother anything – you'll hardly know I'm there."

"Hardly know you're there? We'll see about that."
And so the challenge was given.

There is a collection of pearls in the dream as well, their sheen subtly variant in shape and color. They belong to my foster mother. She tells me they will wait, that we will celebrate together when I win and they will become mine then. I had been visiting with her at her home (a trailer, part of an encampment).

She was very close to me and told me while I was walking away that she wanted to come, too. I greatly disliked the idea of her going into danger with me, but she insisted on coming. I had three heavy things to carry and could only carry two; she came with me and carried the third one.

She thought it would take us about two weeks to get to our

	Waxing Week
Gaiaday	
Sunday	Roman emperor Aurelian instituted games in honor of Sol Invictus on 12/25/274 BCE, to be held every four years from 274 onwards.
Moonday	

destination going at her pace – much farther away than I'd originally believed. I admitted that I didn't know anything about pacing myself and had no idea when or where to stop, rest, eat, or sleep, agreeing that these things could be up to her. I was grateful for her considerable wisdom and experience. An Australian sheep dog followed us part way – one of my father's dogs.

We were masters of appearance. Our encampment was not far from the border. My challenge was to cross over safely while switching the appearance of my gender, and keep them from knowing it was really me. If caught, I'd lose the game and my freedom to cross again would be forfeit forever. If they let me cross first, *then* caught me? Perhaps I would never see home again. Chance was to decide which gender I'd adopt before switching to the opposite. But a new element had been introduced, behind my back and without my consent. The authorities had been alerted to watch for any tiniest trace of the work with illusions for which my people were so well-known. The moment I switched the game would be up and I'd be giving myself away.

Suddenly whole groups of us were on the move, some starting out with their true genders and some with assumed ones but all getting caught as they switched. Each was switching: if one must be caught, all would be caught together. None would be left trapped, stranded, alone. On and on they came, widely varying in their approach. Hidden in bushes near the border, I watched each one to see which approach would win. None did. All were doomed. Finally my foster mother approached the barrier and I rushed up ahead of her as myself, with no time for switching. I fully expected to be caught, but I didn't care, so long as she went free. It was me and me alone they were seeking, and there I was, right before their eyes. But they let me pass! They were looking so hard for traces of illusion that what they really wanted

	Waxing Week
Mercuray	
Vensday	
Tiwsday	Both the soil of Mars and the hemoglobin of human blood are rich in iron; because of this, both are deep red in color.

passed them by. I could still go unmolested across the border and back from the creative, loving culture that was home. No one else was me, so after rigorous testing that proved as much, we *all* went free.

My people rose up as one to save me, their sister, and I gave myself away for my foster mother, but love won, that day. We all went free.

"Le cunamh De agus sin slan..." "With God's help and ourselves healthy..." (in olden days a popular Irish-Gaelic phrase). Aha! With *love's* help. Love saved the day.

Effects of a Story, Told at Christmas

I heard a story once and it was *so* beautiful. I heard it at Christmas in the dark, with candles burning and the light shining beautifully on glass. It is a story they tell in the Navaho Nation, shared with me by a friend who came from there. It was told to me a long time ago, and only once, so I probably have many of the details wrong...

I tried to tell it in my own way, right here, feeling fine about making up a few details if this would let me share the spirit of the story. But that's not happening. When I get to the part that touched me most I remember only what I felt and don't have any details at all. I can't make them *all* up. My details are not theirs; my jokes, too, are informed by a lifestyle so unlike it changes the flavor of anything I could say. So let me just share the barest bones, and simply *tell* you the idea I sensed that touched me so.

It's about a people who were brought into the land by a wizard from Mexico. Normally, they were contained quite harmlessly in

	Waxing Week
Jupitay	
Saturday	
Starday	Ceres had an affinity and joint cult with Terra Mater (Mother Earth). Ovid describes these goddesses as "partners in labour"; Ceres giving cause for crops to grow, and Tellus giving them space for it. Could Ceres be Earth's higher octave?

miniature form behind a small glass, and drawn upon as needed for building or inventing all manner of outlandish things. Sadly, they lost the way back into their glass and are now wandering about in the world, with no one to restrain or guide them. They get into all kinds of trouble on their own, and make even more. What messes they make, and how lost they are!

Listening in the shining darkness, I could clearly see the grandmothers there, comforting the little children. People were being rounded up; people were dying. Kit Carson's atrocities are not fresh in my mind now, but they were then (no one ever mentioned the man; I heard about it elsewhere). I could see the grandmothers, comforting the little children. They know hatred is no way to live. They found this story the children could keep, a story to help them coexist with compassion and not hate. The grace and beauty of the idea brought tears to my eyes.

Horn Dance

In the British village of Abbots Bromley, the people keep a tradition they must hold dear, for they have kept it without fail for many hundreds of years. At earliest written reference (1686) it took place around the winter solstice (though the date was moved in the 19th Century). It may well be older than the written word – the twelve figures in the pageantry look very like certain ones that can be seen in the cave paintings at Lacaux. It's as much dance as it is pageantry: a Horn Dance.

In this village, the participants all came from only one family (a second family name is born by some, acquired by the original bearer marrying in). They have passed the steps and patterns down from generation to generation for as long as anybody can remember. On Horn Dance Day the twelve rise early, and receive a blessing at 7.30 a.m. from the Vicar of the church where the horns are kept.

	Bright Week
Gaiaday	
Sunday	It's not the Sun moving, it's the Earth (even though the Sun is moving too; gravity has us moving in step with it), but from our vantage point, it feels as though we are the ones sitting still. So when it comes to numbering the days that each of the heavenly bodies spends in various signs of the zodiac, it's easier to say it the way it looks to us from here. From our vantage point, the Sun spends exactly one month (30.4 days) in each sign.
Moonday	

Six dancers (Antler Men) will carry huge antlers they will later be holding to their heads. Oddly, these are not deer antlers anymore (if ever they were), but reindeer horns, mounted on wooden deer heads with poles attached to make them easier to carry. The heaviest weighs 25 pounds and the oldest was shown by carbon 14 dating (circa 1970) to be around 900 years old. The dancers used to wear their own clothes, which they decorated with a few ribbons and bits of colored cloth.

Among the other six figures are a Dancing Fool who carries "what looks like an inflated pig's bladder" (Ravenscroft) around on a stick; he will spend as much time chasing folks with it as dancing. There is also an Archer with bow and arrow and another man, dressed as a woman (Maid Marion). A fourth man wears the Hobby-Horse, and to complete the ensemble, there are two musicians – an accordion player and a boy with a triangle. They play traditional English folk tunes. In earlier days the tunes were different, and the favored instrument was a fiddle; earlier still, tunes were played with pipe and tabor.

Having been blessed, the participants set off, bound for twelve locations in what was once a twenty-mile radius (today though, it is only ten), to visit many of the local houses and farms. If they don't show up at your place and dance for you there, you miss out on the luck they bring for the season, "so their arrival is always greeted with relief. The dance itself begins with the horn-bearers leading the way into the garden or the farmyard, where they form a silent circle. The music begins and the leader suddenly breaks the circle, passing between the second and third of the dancers. The rest follow him, then form two lines with three dancers in each line, facing each other. They raise their antlers as if they are about to fight, then back away, advance, lock horns..." (Ravenscroft) ...and always the one with the triangle keeps time, and the Hobby-Horse snaps his jaws to that rhythm.

	Bright Week
Mercuray	Mercury is associated with travelers, short trips, and relationships with those you see most often.
Vensday	
Tiwsday	

Winter Moon

The day is still celebrated, not on the winter solstice but on the first Monday after the first Sunday after the fourth day of October, and much is made of it. There are food stalls, crafts, and exhibitions. The town itself is lovely.

I take this information from John Ravenscroft's amusing article, offering more than I include here, though I *have* relayed all that speaks of the oldest ways (for more, see Back Pages, pp. 274-6).

Our Christmas Surprise

*"No stockings this year – I can't make it home,
or beleaguered Gladys would be left all alone;
perhaps we three can build some Christmas cheer.
No. This year I really must have Christmas here."
But sly old Santa, that ingenious elf
(he always takes so much on himself)
with a hee hee hee and a ho ho ho,
he was up to something I couldn't know.*

*A mysterious box, all festive and gay
appeared in our mailbox one unsuspecting day.
We snipped off the paper and whisked off the lid,
the better to see what lay there hid.
Our Christmas stockings must have flown through the skies,
for their contents were here, before our eyes!
There was the darlingest pear – it could tell time,
always and forever would it be in its prime.
With foot, hair and body care in lavender and blue
our stockings had found us, it had to be true!
With us on the couch and the box on my lap
It was hard to decide which first to unwrap.*

	Bright Week
Jupitay	
Saturday	
Starday	Comets and novae have been observed and discussed for several thousand years. Comets in particular were portents of great interest to ancient people, much puzzled over and quite variously interpreted.

Eagerly unwrapping one by one
I pulled things out with James looking on.
Then came bags of goodies – a glorious sight!
James' eyes, especially, were all alight.
"I was really hoping for something to eat,
and they have given me such a treat!
'Non pareil' almonds – that means 'nothing but the best!'
They must really like me. I'm so impressed!"

There were cashews and hazelnuts and sunflower seeds –
delicious things that James really needs.
And I like them too. That generous man
said I'm welcome to eat them – all that I can,
even though I know they're his favorite snack.
For kindness and fun, he has such a knack!
Oh, we really loved our Christmas surprise –
it tickled our toes and twinkled our eyes!

Recipe for a Christmas Cookie

I took an alternate route to a traditional cookie and discovered a new variety that's ridiculously easy to make; I could *not* leave them alone. My daughter (a baker by hobby) called them addictive. When we made them it was high summer. We had just bought some unrefined virgin coconut oil. It had melted, and smelled amazing. We wanted a buttery cookie and I had a shortbread recipe right there, so I whipped it out, stirred flour, sugar, and salt together, poured the liquid coconut oil (without having to melt butter) into a different bowl and stirred in some flavorings, then poured the mix into the bowl with the flour. It took only moments to do. Turn the page to see how I did it, and feel free to substitute the flavorings I used with any *you* prefer: lemon and lavender, maybe cinnamon in the flour bowl or sprinkled on top.... Have fun!

	Waning Week
Gaiaday	
Sunday	The Southern or Northern Lights shine out when charged ions spinning off from the Sun get trapped in the Earth's magnetic field and smash into air molecules above the poles; making them glow in shades of green, blue, pink and red fifty to four hundred miles above the Earth: a spectacular sky show.
Moonday	

Winter Moon
California Dreamers' Not-Quite-Shortbread Cookies

Take:

1 cup flour 1/3 cup sugar and ¼ tsp salt 1 tsp vanilla flavoring ½ tsp lemon flavoring ½ tsp orange flavoring 1/3 cup coconut oil	*Place first three ingredients* *in larger bowl and combine well.* *Pour remaining ingredients* *in smaller bowl, stir well, and* *pour into larger bowl. Stir together* *until completely blended.*

Preheat oven to 325 degrees. Drop dough by rounded tablespoons on ungreased cookie sheet. Flatten and shape as desired (I cut crescent moons with a bottle cap in some and dug simple designs with a toothpick in others). Bake 25-35 minutes (until lightly brown). Let cool for 10. Makes 12-15 cookies.

I used bread flour (what I had on hand; this may not matter much). If oil is not fresh-bought or protected in the fridge, the cookies won't have that amazing flavor until they have sat for several weeks.

The Word 'Tuesday'

Do you think you know by now where the word 'Tuesday' comes from? I did; I thought it was drawn from Tyr, the Teutonic god of war, but I was mistaken. In truth, Tyr of the Vikings had obscured Tiw of the much earlier Anglish/Anglo Saxons, at least in *my* mind. And it is actually Tiw for whom Tuesday is named. From Ms. Heath's meticulous, scholarly, and excellent article (see Back Pages, page 264) I learn: "According to Simek, Tiw's name was derived originally from proto-Germanic *Tiwaz:* 'god of war, the skies, and the Thing'." The Thing was really many things, among them a gathering not only of individuals but of

	Waning Week
Mercuray	
Vensday	
Tiwsday	Like Jupiter, Mars was once a storm deity who thundered.

tribes or clans. It was also a time, and a place, where legal disputes could at last be settled. It could be a place of reconciliation as well as justice.

Those who defended Hadrian's Wall in Rome's service were drawn from a mix of subject nationalities. Among them were a group of Frisians, who left a temple in honor of their Mars Thincsus (for the Romans had equated Tiw with Mars). Saussaye opines in his *"Religion of the Teutons"* "It appears likely that the Frisian cavalrymen, who call themselves citizens, saw in Tiu the god not only of the squadron but also of their popular assembly, the Thing," and in another work discussing same, Anne Ross observes that "Mars Thincsus would seem to be the only non-Roman god to figure as an orthodox Roman warrior" *(LOL)*.

In Ms. Heath's article, I learned so much, including the fact that horses were special to the Anglish. They were used in some way for better communication with the world of spirit. There is even a hill in Warwickshire sporting a horse design, cut deep into the red clay, now known as "the Vale of the Red Horse."

No other deity's name has become part of any runic alphabet at all, despite Odin's claim to have invented them, yet one finds Tiwaz among the Norse runes. "His symbol was an arrow pointing up; to see one was thought to be..." a message of support "from the god himself. His very name was power; when cut into one's spear, the weapon became a talisman of protection." From Thorsson I learn that "The rune refers to justice or order or law, together with the self-sacrifice of one dedicated to preserving the same," and that Tiwaz has a third meaning, as deeply intertwined as the first two. It is the "world column," preserving order by separating heaven from earth, and thereby allowing manifestation to happen freely "in the created space." According to him, it was thought to be "the sky pole," marked by the North Star. In a certain Old English Rune Poem, Tiw actually is described as "a guiding star" (1). Apparently, the Anglo-Saxon kings of Essex even claimed him as an ancestor (2). Evidence from an early

	Waning Week
Jupitay	
Saturday	
Starday	This is Starday of Waning Week, or Waning Starday, the last of any given lunation and best for saying goodbye. The second half of the Crone Moon Meditation is meant to happen here. Sit at peace in the dark and notice all you have let go. Is anything still stuck? Breathe it out – feel it go. And breathe in peace. Celebrate space. For more on these lunar meditations and the lady who put them together, please see my Back Pages (p. 277).

runic script dating back to the 2nd Century depicts Tiu as "the most venerated god" the Germans had (3). I learn from Ms. Heath that the word "Tiwaz" is as much a derivative of the word "Dyeus Pter" as Zeus and Jupiter are, and wonder what became of him. The Anglish were Christianized so much earlier than the Vikings. Perhaps this is why we know so little about them. For sources to which numbers above refer, see Back Pages (p. 264).

How Nordic Yule Was Moved

The Abbots Bromley Horn Dance is not the only celebration whose date has at one time been moved. At Christian insistence Yule was moved, to the winter solstice on the Julian calendar, December 25th by its own reckoning (which explains why Bede believed it began after sundown on Christmas Eve). The first recorded date of Christmas being celebrated on that day was in 336 AD, during the time of the Roman Emperor Constantine. Before Yule was moved, it was celebrated on the first full moon following the winter solstice, and it lasted for three days and nights (for more, see my Back Pages, p. 277).

> *"The rising of the sun, and the running of the deer,*
> *and the playing of the merry organ, sweet singing in the choir.*
> *The holly bears a blossom as white as lily flower,*
> *and Mary bore dear Jesus Christ to be our sweet savior..."*

I've loved that Christmas carol ever since I was small, probably for the way it resonates with a feeling inside me, about the deer, and the sun, and Christmas time. Why that should be was always a mystery, but I loved the way it made me feel. It seems to cherish the love of God however it is expressed, whether in forms old or new.

I went online to make sure I had the lyrics right – and found quite the surprise: a game! For more about it, see Back Pages, p. 278.

Thirteenth Lunation

Where bliss may arise.

241

Dark Moon

Riding the skies
     between formless and form,
          between born and unborn in the
velvet night,
   an empty moon in the sun's arms lying –
        deep in stillness, scrying
     that silent tune –
            gathers seeds of light.

Her mystical form,
      from unborn to born
          in sweet darkness lies,
    where bliss may arise.

	Waxing Week
Gaiaday	
Sunday	In Rome, the Philocalian calendar of AD 354 names a new festival, "Natalis Invicti" (birth of Sol Invicti: the Unconquered Sun, god of soldiers), to be held on December 25th.
Moonday	

How Our Current Calendar Came To Be

Calendars are places where a group's story meets in time. It may be worth noticing – what stories have *we* been telling? We inherited our current calendar from Rome. It was begun by Romulus in 753 BCE. Initially it had but 10 months, from March to December. For the uncertain time of winter two new ones were added, shortly after the end of Romulus' reign: January for Janus (god of endings and beginnings) and February for Februa (meaning "to purify"). The fifth and sixth months were called "Quintilis" and "Sextilis," names as faceless as September (meaning "seventh one").

The first day of each month was dedicated to Juno and sacred to Janus. Debtors had to pay off their debts on this day. Initially, Roman months were identical to the lunar cycle. Each was divided into sections corresponding to the first three phases of the moon, and all days were referred to in terms of one of these three phase names: Kalends, Nones or Ides. A "pontifex" (another word for priest) was assigned to observe the sky. When he first sighted a thin lunar crescent he called out that there was a new moon and declared that the next month had started. For centuries afterward, Romans referred to the first day of each month as Kalendae or Kalends (so March first would be "the Kalends of March"), from the Latin word calare (to declare). The word "calendar," too, derives from this custom.

Nones (from Latin nonus, meaning "ninth") identified the day when the moon reached first-quarter phase. When the pontifex first saw the lunar crescent he noted its width and with practiced knowledge, figured the number of days expected to pass between then and first quarter, stating it right after announcing the new crescent. If he called out "six," the day following Kalends would be called the Sixth of the Nones of (month's name here). If that month were March, this would be our March 2nd, six days before the Nones (first quarter) and one day after the Kalends (the new

	Waxing Week
Mercuray	Mercury is the ruling planet of both Virgo and Gemini. It moves fast, spending only 7.33 days in each sign.
Vensday	
Tiwsday	

moon of March). It would be followed by the Fifth, then the Fourth and so on, until Pridie (Latin for "the evening before"). Any day preceding a new phase was "Pridie ___." In this case, it would be "Pridie Nones of March," the day before first quarter: March 6th, in our thinking. And if "six" *were* the number the priest called out, Nones and Ides would fall on the seventh and fifteenth days of the month, but if "four," the Nones would be on the fifth and the Ides on the thirteenth.

Despite this variability, the difference between Nones and Ides was always eight, as implied by the very word "Nones" ("ninth"), for eight days prior to the new phase plus the day of the new phase itself equals nine. The actual number of days between lunar phases varies between eight and seven, averaging out to 7.4 (and lunations as a whole average 29.53) days each, but Romans didn't like even numbers, so they went for "Nones" instead of some other word ("Octets" perhaps) based on eight. A full moon was called "the Ides," and all were dedicated to Jupiter. After the Ides had passed, the countdown to the next cycle encompassed either 15 or 17 days, varying on account of constant changes in positioning between earth, sun, and moon.

By the 1st Century BCE, the calendar had become hopelessly confused. Its year stayed in step with the seasons through an extra cycle inserted when called for, but real confusion was brought on by political maneuvers, for the Pontifex Maximus and College of Pontiffs had the authority to alter the calendar, and this they sometimes did, thus reducing or extending the terms of certain magistrates and similar public officials at will.

In 45 BC Julius Caesar put an end to it all by reforming the Roman calendar completely. He also moved the celebration of New Year's Day from March to January, thus favoring Janus over Mars. On this day, Romans exchanged gifts and promised to be better with each other in the year to come, throwing parties,

	Waxing Week
Jupitay	Jupiter is the ruling planet of Sagittarius and Pisces and is exalted in Cancer. Taking 11.9 years to orbit the Sun, it spends 361 days (most of an Earth year) in each sign.
Saturday	
Starday	

eating,drinking, and dancing. Month lengths were extended at
this time to their modern lengths too. Rome appreciated their
Caesar and his new design so much that in 44 BCE
Quintilis (Julius' birth month) was renamed Iulius, in his honor.
Years later in 8 BCE, Sextilis was renamed August, for Augustus.
The month name April is thought to come from "aperire"
(meaning to open, like buds in spring), May is from Maia (Atlas'
daughter), and June from Juno (Jupiter's wife).

Our Gregorian calendar was a reform of this Julian one,
instituted by papal bull in February of 1582 by Pope Gregory
XIII, for whom the calendar is named. The motivation for this
adjustment was to return the date for the celebration of Easter to
the time of year it was celebrated in when the early Church
introduced it. The reform was initially adopted by the Catholic
countries of Europe and their overseas possessions, but over the
next three centuries the Protestant and Eastern Orthodox
countries also moved to what they called the Improved Calendar,
though Sweden did not adopt the new style until 1753. Russia
kept to the old way even longer. This is why the Russian
Revolution of 1917 is called the "October Revolution," even
though it happened in Gregorian November. It finally adopted
the new style in February of 1918, but Greece kept to the old way
until 1923.

So this is where the calendar we use today comes from and
how it came to us in so unified a way: under Rome to a world
united through military conquest. When too depleted to remain
they withdrew, too late to bring any joy to the populaces they had
controlled; for Rome had depleted its occupied places to extend
its own time, costing native peoples their edge and their ability to
defend themselves. The eagle (Jupiter's primary sacred animal)
had become one of the most common symbols of the Roman
army. Folks may have lamented that the eagle had flown, yet
even today, we see the footprints of that symbolism in Europe

	Bright Week
Gaiaday	
Sunday	
Moonday	The Moon is the ruling planet of Cancer and is exalted in Taurus. It moves fastest, spending only two and a half days in each sign.

and even in the U.S. Such was the Rome of which our Western World is child (or in some cases, grandchild).
I learned of Rome's lunar calendar here:
http://www.webexhibits.org/calendars/calendar-roman.html

A Calendar of the Anglish

It wasn't as though the Anglish didn't have a calendar of their own before Rome intervened. I learn from Alkman's blog, Mine Wyrtruman's site and Paul Anthony Jones' *Mental Floss* that the only description we have of the Anglo-Saxon Heathen (i.e. pre-Christian) calendar comes from *De Temporum Ratione (On the Reckoning of Time)*, which was written by the Venerable Bede in 725 CE. Here, he relates the calendars of several ancient cultures, giving us a lot of information about the Anglish and their year that may otherwise have been lost, like the names of the months, the reasoning behind them, and insights into how observing the skies set the framework within which their idea of time moved. The calendar itself was luni-solar. Here are the pre-Christianized month names, alongside our current ones:

Æfterra Geola = *January*   Solmonath = *February*
Hrethmonath = *March*   Eostremonath = *April*
Thrimilci = *May*   Ærra Litha = *June*
Æfterra Litha = *July*   Weodmonath = *August*
Haligmonath = *September*   Winterfylleth = *October*
Blotmonath = *November*   Ærra Geola = *December*

They had but two seasons (winter and summer), each six months long and peaking at the solstices. Their year looked like this:

	Bright Week
Mercuray	
Vensday	Venus is the traditional ruling planet of both Libra and Taurus. It is exalted in Pisces, but by ancient reckonings took joy in Leo, and spends about 18.75 days in each sign.
Tiwsday	

Weavedancer Moon

Winter Months: Winterfylleth, Blotmonath, ÆrraGeola, (Winter Solstice)ÆfterraGeola, Solmonath, Hrethmonath;
Summer Months: Eostremonath, Thrimilci, Ærra Litha, (Summer Solstice)Æfterra Litha, Weodmonath, Haligmonath

Bede says their months were lunar (twelve plus the occasional thirteenth), but he doesn't say which lunar phase they started on. From Robert Sass I learn that the Saxon month began on the new moon, and as with the Anglish, their day began at sundown. The languages and worlds of the Saxon peoples differed from the Anglish in many ways, yet many things they shared. We cannot be certain about which phase began the Anglish months, but we do know each seasonal marker was on a full moon (this was the method of dating for all Germanic tribes). In the Anglish calendar three of these four were holidays, lasting three nights each.

The first was winter's beginning, in the Moon of Winterfylleth. Just *see* that moon growing full: winter filleth, indeed it does! The second was mid-winter (Geola, or Yule); in pre-Christian days it was a three-night holiday that took place on Æfterra Geola's full. "At Yule it was determined if a thirteenth moon would be added to the year...to keep the following year's Yule as the first full moon after the first new moon after the solstice..." (Andreas Nordberg, Norse scholar). The origin of the word Geola" is obscure, but it may mean or be related to the word "wheel" (Alkman). This actually makes sense. If the wheel in question is the wheel of the year, the reference that sets where the year begins would be well named "wheel" (the very place within which that wheel begins its spin). Bede tells us Yule refers to "the day the sun turns around," or days begin to lengthen (Wyrtruman).

That third holiday was the First Days of Summer. All Germanic Heathens venerated the same full moon as the start of

	Bright Week
Jupitay	
Saturday	Saturn is the ruling planet of both Capricorn and Aquarius, and is exalted in Libra. It takes 29.5 years to orbit the Sun, spending about 2.46 years in each sign.
Starday	

253

summer; the Anglish and Franks called this Eostre. The Church doesn't date Easter to the equinox, but to "the first Sunday after the first full moon after the vernal equinox..." (said Bede), and it does so in the same way to this very day. This pre-existing full-moon tradition may explain why. Midsummer had no festival. Instead, it would sometimes get a whole extra lunation. They would decide at Yule whether to add an extra lunation to summer-time or not, and in this way keep the moons in step with the seasons.

Bede does say a bit more about them. Want to explore? ÆrraGeola means "before Yule" and ÆfterraGeola means after. The solstice was a useful reference. I can see why it would be book-ended this way.

The next moon is Solmonath, a name that's puzzled many. During this moon the Anglii gave savory cakes and loaves of bread to their gods, hoping for a good harvest (Jones). But no word "sol" with "cake" attached as its meaning exists in the Old English language (Alkman). There *is* a word *sōl*, but it means "wet sand" (Jones) or "mud" (Wyrtruman). I doubt that they offered their divinities mud pies! Perhaps these cakes had a mud-like color or a sandy, gritty texture? Or sol could be a borrowed word meaning sun (Alkman). But why borrow a word when you have your own unless the tradition were borrowed too? These cakes may have been symbolically sun-shaped. I have heard of so(u)l-cakes before and so have you. We've even got the recipe! It puts me in mind of those King Cakes eaten in New Orleans, during this very lunation (Solmonath, which is February).

Next is Hrethe Moon, named for Hretha. Apart from the fact that her name means "glory" we know next to nothing about her. Could she be the one who battles winter, creating space for the new life that follows? Summer begins at the next moon's full, so maybe (Wyrtruman). To Eostre (a goddess of the dawn) the people sacrificed next. Her moon follows. As well, there was

	Waning Week
Gaiaday	Earth's magnetic pole keeps moving. Since tracking began in the 1830's, the North Pole moves 24 miles to the north each year. It will (probably) leave the U.S. for Siberia in another couple of decades.
Sunday	
Moonday	

her festival, following the spring equinox. Bede tells us no more. Next we have Thrimilci ("three milkings"), when the cattle fed so well on all that fresh spring grass that they needed to be milked three times each day.

Ærra and ÆfterraLitha follow. The summer solstice falls between, and summer could be seven months long if a leap moon was needed. Years with such summers were called Thrilitha ("three Lithas"). The word "Litha" may mean "moon"(Alkman). Since this is where leap moons went, that could make sense. Then again, it may mean gentle or mild, and refer to the weather and the seas (Wyrtruman), for summertime was Boating Season.

Weod-monath is next. It means "Weed," or "Tare," or "Plant Moon," "for they are plentiful then." Some weeds can be tasty, and some have medicinal value. I wouldn't name a moon after a useless thing. Would you? The next moon is Halig-monath, meaning Holy Moon: a time of offerings. Many cultures were enjoying their harvests about now and offering up the fruits of their fields. Though Bede says nothing, I bet this happened here, too. Winterfylleth is the moon that follows. It speaks for itself.

The last moon is Blotmonath, which may mean Blood or Sacrifice Moon. Bede doesn't say much, but the cold has come. Pasturage is scarce, and possibly other foods as well. It's time to guess how many cows you can keep alive through the winter, and how much frozen dead cow you'll need for yourself! I'm guessing a lot of slaughtering happened then, with many prayers.

"Use of the Germanic calendar dwindled as Christianity, bringing with it the Roman Julian calendar, spread more widely across England in the Early Middle Ages. It quickly became the standard; in Bede's day he could dismiss the Germanic one as the product of an 'olden time'"(Wyrtruman).

	Waning Week
Mercuray	
Vensday	
Tiwsday	Mars is the primary ruling planet of Aries and is exalted in Capricorn. It orbits the Sun in 687 days, spending about 57.25 days in each sign.

Weavedancer Moon
Concluding Thoughts

I've been dreaming here. And one thing in particular has become clearer to me: when we truly notice things, as they happen, close to where we are, we may just possibly see a living world more magical than any *we* could have imagined. I believed but did not see this so well, until it was reflected back to me in the writing process that produced this book.

My whole focus, originally, had been simply to recreate the old luni-solar calendar, because I wanted to see what it felt like. I learned a lot, but never could catch the feeling for myself. I have, however, produced luni-solar calendars calibrated for current years, both for North America's West Coast and along the Greenwich Meridian. I don't know whether there would ever be any interest in such a thing or not. If there is, let me know by emailing 13lunations@gmail.com, and I'll see what I can do about making it more widely available.

Yes, I've been dreaming here. You can see the results. We can dream anything at all, if that's what we want. The names for the months and the mythos for astrology (the planets and the zodiac figures – constellations that are really only groupings of stars) all come from those who conquered. What were the original dreams? What would the local dreams of a calendar of local place look like? We carry the vision (the memory) for this ourselves. You've seen emerging from me here mostly Celtic and Teutonic things, but that's because *I'm* mostly Celto-Teutonic. How about you? Where does *your* family come from? Your dreams know. They can take you places the same way mine have done. Awareness is in all, and the possible is without limits. If you're interested in any of this, check out the "Indigenous Mind" program offered in Oakland, CA at Naropa University. Using self as map, with genealogy, history and one's own dreams as guideposts, people learn and do the most amazing things there. ancestralapothecaryschool.com or www.traditionalknowledge.org can tell you more.

	Waning Week
Jupitay	
Saturday	
Starday	Neptune spends 13.75 years in each zodiacal sign, Uranus only seven, Pluto is highly variable (spending between 15 and 26 years in each) and Ceres moves through approximately three signs every year.

Weavedancer Moon

Oh, and I thought I'd better mention one more thing: any poems you see scattered around here with no author attached are mine. I couldn't resist plopping them in.
Well, this dreaming-time is over. Good morning, everyone!

Back Pages
(Acknowledgements, Notes, and References)

Many people helped me write this book. I'd be rambling on and get struck by an idea, then wonder whether there was any truth to it or not. Off to the internet I'd go, where so many fine people had shared such interesting things. Fully expecting to acknowledge them all, I began my studies, getting so excited by the things I learned there and so deep in the process of figuring out how (or even if!) they integrated with my own ideas that by the time I was ready to start adding references (in the beginning of the long process that produced this book especially) I would find I had accidentally erased them. Often months (and in at least one case years) had gone by, so search as I might, I could not find them again. I hope these internet contributors can forgive me. I just want to say thank you – I have learned so very much from you all.

I would also like to thank my mother, who shares my love of this living earth, and Bob, who witnessed much of what I share here. Thank you Teresa, for being my sister in writing. And Brittany, it was your observations that inspired the more scientific aspects of this book. James, especially – thank you. You know why.

More on the Nine-day Week

When I searched for information to support the concept of our week's daily flow having once been in a different order or explain when or why that order had been changed, what surfaced instead was a weird mathematical argument used to "demonstrate" how the current order is how it always has been and how it always should be, together with an article that tried to claim biblical support for the same idea. One would not expect any of the Bible's authors to see value in days named after Babylonian gods as seen in "wandering stars," unless a few were Babylonians

261

themselves.

I next asked questions about the Judaic calendar, looking to corroborate the idea that Judaism once made use of a nine-day week, but the search engine encouraged me to ask about the Hebrew calendar instead. When I consented, an article surfaced noting a new argument among certain modern Jews for their day being meant to begin at sunrise and not sunset after all, but from that point on, no references to the Judaic calendar ever surfaced however diligently I searched, only those for the Hebrew one. Other articles did arise that were suitable to our search, both of them from Wikipedia:

As a sub-article under "Celtic Calendar":

"Medieval Irish and Welsh calendars[edit]
Further information: Gaelic calendar and Welsh holidays

Among the Insular Celts, the year was divided into a light half and a dark half. As the day was seen as beginning at sunset, so the year was seen as beginning with the arrival of the darkness, at Calan Gaeaf /Samhain (around 1 November in the modern calendar).[4] The light half of the year started at Calan Haf/Bealtaine (around 1 May, modern calendar)...The Laws of Hywel Dda (in editions surviving from the 12th and 13th centuries) make repeated references to periods of *nine days* (*nawfed dydd*), rather than the "eight nights" that make up the current word *wythnos* *.[7]"

*That miniature number 7 (above) is referring to this, the 7th reference: *Wade-Evans, Arthur (1909). Welsh Medieval Laws. Oxford University Press. Retrieved 31 January 2013.*

Also:

"Lithuanian calendar

From Wikipedia, the free encyclopedia
...*The Gediminas Sceptre, (Gediminas was Grand Duke of Lithuania, from 1315/1316 – 1341, the year of his death)

discovered in 1680, indicates that during his reign the year started in April and was divided into 12 months, varying in length from 29 to 31 days. Each month began with a new moon; the weeks were nine days long [2]* "

* Libertas Klima. _"Zmogus ir gamta etninėje kultûroje"_ (PDF). Lietuvos kultūros darbuotojų tobulinimosi centras. Retrieved 2009-01-21.

Bookish
(Quick References
Not Fitting Within Lunations)

Third Lunation: the phrase "open-eyed watcher of the skies" (p. 50) and the concept of the Teutonic tribes' feelings about newness (p. 54, second paragraph) are both from Gerald S. Hawkins' book: _Stone Henge Decoded_, p. 143, third paragraph up from bottom.

Paschal Eggs (p. 52) are drawn from Madeleine Pelner Cosman's book, _Medieval Holidays and Festivals: A Calendar of Celebrations,_ p. 43 under "Pace egging" (published in 1981).

Sixth Lunation: "British royals had strong connections with the deer..." (bottom of p. 102) contains concepts I absorbed from Marion Zimmer Bradley's _The Mists of Avalon._ While (as she herself has said) "any attempt at re-capturing the pre-Christian religion of the British Isles has been made conjectural by the determined efforts of their successors to extinguish all such traces," I, like her, am aware that even scholarly opinions differ, sometimes sharply; these concepts may have had to be picked out from varying or even contradictory sources. Victors are the ones who write the stories. Sometimes we see better when we look with our hearts.

Ninth Lunation: "The Constellations of Virgo, Libra, and Scorpio" (pp. 178-180) is drawn from these sources (and when quoted, marked correspondingly):

(1) www.zodiac-love-compatibility.net/virgo/virgo,
(2) starsignstyle.com/libra-the-scales,
(3) www.crystalinks.com/virgo.html,
(4) www.space.com/2495-scales-scorpion.html, and
(5) http://www.gods-and-monsters.com/

Twelfth Lunation, "Le Cunamh De": I'd like to acknowledge Patricia Pugnier. Her family taught Irish for many years. They made it such fun! She and her brother have a book out through Amazon (Speak Irish Now) that recaptures the basics very well.

The Word 'Tuesday' (pp. 238-240):
When a named source is quoted I'm drawing from Catherine Heath's *One Hand or Tiw?* at https://greatvalleykindred.com, (with the exception of Edred Thorsson – I found him myself). Towards the bottom, to save space on that page I inserted numbers, knowing I could include them here. Showing them below as contextual quotes.

1: "...Tiw is described as 'a guiding star' in the Old English Rune Poem (Dickins, Bruce; Runic and Heroic Poems of the Old Teutonic Peoples)"

2: "Dr Ellis Davidson explains this inclusion of the Tiw rune on cremation urns by theorizing that Tiw and the god Seaxnet were one and the same, and that he was claimed as an ancestor by Anglo-Saxon kings of Essex."

3: "As Enright points out, evidence from the early runic script that dates back to around the 2nd Century shows Tiu to have been the most venerated god by the Germans."

More on Walpurgis Nacht (Fifth Lunation)

"Though conquering tribes imposed a contrary faith..." These were Charlemagne's tribes. He was a Frankish king living between 742 and 814 CE, who was born near the town of Liège (in modern day Belgium). So ambitious was he that he declared war on many neighboring kingdoms, invading Saxony in 772 and eventually achieving its total conquest.

His primary goal was to unite all of Western Europe under his power and then convert all the Germanic people to Christianity. He persuaded many eminent scholars to come to his court, establishing a new library of Christian and classical works. He introduced and initiated many important legal and administrative reforms throughout his lands as well, standardizing weights, measures and customs dues (http://www.bbc.co.uk/history/historic_figures/charlemagne.shtml). His major accomplishments include developing the rules of the feudal system, encouraging reading and writing throughout his empire, developing commerce with a unified monetary system, and the unification of all Germanic peoples into a single, Christian kingdom (https://www.reference.com/history/major-accomplishments-charlemagne), effectively bringing an end to the Migration Period (476-800: a name that refers to the migratory movements of a wide variety of peoples, including the Huns, Goths, Bulgars, Alani, Suebi, and Franks).

In their entirety his territories became known as the Carolingian Empire, which endured, not giving way until the 1800's (during the Napoleonic Wars). It was later called the Roman Empire and later still the Holy Roman Empire. At its height the empire's territories included Germany, Switzerland, Liechtenstein, Luxembourg, the Czech Republic, Slovenia, Austria, Croatia, Belgium, the Nether- lands, and parts of Italy, France, and Poland.

I learned about Walpurgis Night almost exclusively from an

article a woman named Karen wrote, shared on her website
https://germangirlinamerica.com/what-is-walpurgisnacht/. My
three numbered references come from there:

1: "In villages and towns, people decided that noise would scare
   the witches away, so beginning at sunset, men and boys
   would do all they could to make noise all night long...
   banging pots and shooting pistols into the air."

2: "After her death, the walls of her tomb began oozing a
   healing oil. Because of this miracle, the church canonized
   Walpurga...Seems like Her feast day, is May 1, and she is
   considered the Patron Saint of Coughs, Storms,
   Hydrophobia and Sailors."

3: "Ironically, in another twist of Pagan custom, children would
   gather greenery from Juniper, Hawthorn, Ash and Elder
   trees, and hang it around the house and barn. Once upon a
   time these were considered offerings to the goddess, now
   they were used to frighten witches and other evil spirits."

More on Water (Seventh Lunation)

Jaques Beneveniste (p. 132) published his paper in the summer
of 1988, in a scientific journal by the name of *Nature*. William
Reville, professor of biochemistry and public awareness of
science officer at UCC discusses some of Beneveniste's ideas in
an online article published as part of *The Irish Times* (find him at
understandingscience.ucc.ie). You can find more about
Beneveniste in "Could Water Really Have a Memory?" (written
by Simon Singh, co-author of *Trick or Treatment? Alternative
Medicine on Trial)* where the author weighs in with similar views.
You will, however, get a very lopsided picture unless you also see
Water Memory, a 2014 documentary about Nobel Prize laureate
Luc Montagnier, found on Youtube. The video revisits Jaques
Beneveniste's work as well.

It became clear to me while exploring this that the web caters to whatever orientation a person appears (by its own peculiar yard stick) to lean toward. My search on two different browsers, one belonging to me and the other to a person whose search patterns were quite dissimilar made this appallingly clear. These algorithms keep us all in our own little bubbles and, in my opinion, are what do the most to divide us from one another.

The Legends of St. Senara, A Blended Tale (10th)

Senara is the patron saint of Zennor, but little is known of her historically. Legend connects her with Princess Asenora of Brittany, who married King Goello, and in Cornish folklore she is Azenor, the mother of their Saint Budoc.

In one story, "Senara was a princess from Brest (in Brittany) originally named Asenora, who had 'a rather dubious reputation' before her conversion. Her husband the king wrongly accused her of adultery and she was condemned to be burnt. When it was discovered she was with child, the gaolers nailed her into a barrel and threw it into the sea, to avoid being guilty of murdering the unborn child. She was miraculously fed by an angel, bore her child in the barrel and was washed up on the shores of Ireland."

In another tale it was a stepmother who, jealous of Asenora's beauty and virtue, made the accusations and insinuations that got her and her barrel thrown into the sea. While off the westernmost end of Cornwall she gave birth to a son in the waves, who later became either Saint Budoc or some other (Irish) bishop. She then washed up on the Cornish coast and founded a village to which she gave her name - Zennor - before continuing to Ireland.

There is a third story that only knows her as "a Christian princess from a far off land. When barbarians chased her to a cliff she prayed to God for safety and she jumped, but rather than being crushed by the fierce waves or impaled on the rocks below God gave her a fish's tail, and she swam to safety through the worst storm ever created since the flood. After she converted the town

268

that bears her name, she jumped back into the waters to keep an eye on her village from the fathoms below."

But the evolution of this mermaid does not stop here. In the following tale she has no name but has connections still to the church in which her chair rests:

"Hundreds of years ago a very beautiful and richly attired lady attended service in Zennor Church occasionally – now and then she went to Morvah also; her visits were by no means regular – often long intervals would elapse between them. Yet whenever she came the people were enchanted with her good looks and sweet singing. Although Zennor folks were remarkable for their fine psalmody, she excelled them all; and they wondered how, after the scores of years that they had seen her, she continued to look so young and fair.

No one knew whence she came nor whither she went; yet many watched her as far as they could see from Tregarthen Hill. She took some notice of a fine young man, called Mathey Trewella, who was the best singer in the parish. He once followed her, but he never returned; after that she was never more seen in Zennor Church, and it might not have been known to this day who or what she was but for the merest accident.

Early one Sunday morning while our mermaid was out on the ocean a-fishing, a vessel cast anchor about a mile from Pendower Cove; in due time she came up close alongside and hailed the ship. Rising out of the water as far as her waist with her yellow hair a-floating, she requested the captain to trip his anchor just for a minute, as the flukes of it rested on the door of her dwelling, and she was anxious to get in to her children to dress them and be ready in time for church. Her polite request had a magical effect upon the sailors, for they immediately "worked with a will," hove anchor and set sail, not wishing to remain a moment longer than they could help near her habitation. Sea-faring men, who understood most about mermaids, regarded their appearance as a token that bad luck was near at hand.

When Zennor folks learnt that a mermaid dwelt near Pendower, and what she had told the captain, they concluded – it was this sea-lady who had visited their church, and enticed Trewella to her abode. To commemorate these somewhat unusual events they had the figure she bore – when in her ocean home – carved in holy oak, which may still be seen."

A more modern version puts things this way:
"Matthew Trewhella was a good-looking young man with a good voice. Each evening Matthew would sing the closing hymn at the church in Zennor, solo. A mermaid living in neighboring Pendour Cove was enchanted by the music. She dressed in a long dress to hide her long tail and walked a bit awkwardly to the church. Initially, she just marveled at Matthew's singing before slipping away to return to the sea. She came every day, and eventually became bolder, staying longer. It was on one of these visits that her gaze met Matthew's, and they fell in love.

However, the mermaid knew she had to go back to the sea or die. As she prepared to leave, Matthew said 'Please do not leave, who are you, where are you from?' The mermaid told him that she was a creature from the sea and that she must go back. Matthew was so love-struck that he swore he would follow her wherever she went. Matthew carried her to the cove and followed her beneath the waves, never to be seen again.

It is said that if you sit above Pendour Cove at sunset on a fine summer evening you might hear Matthew still, singing faintly on the breeze."

The above collection of stories were found on these websites:
www.britainexpress.com/counties/cornwall/churches/zennor.htm
and https://en.wikipedia.org › wiki › Saint_Senara

Senara was highly venerated by local fishermen and is said to represent the dual nature of Christ (human and divine). Medieval folk regarded her as a symbol of lust and a warning against the sins of the flesh, due to her story of purported adultery and subsequent conversion, and the Irish say that mermaids are old pagan women transformed and banished from the earth by St. Patrick. "Though they spend most of their time underwater, they have been known to assume human form and come ashore to markets and fairs."

Here is my somewhat fanciful take on the above patchwork of tales. They depict the evolution of a particular soul. (Do whole cultures have souls?) I see the Celtic princess Asonora zestfully partaking of what was right for her culture; and when that culture was conquered (or converted, or whatever the case may be), she had the grace and flexibility to evolve, partaking with equal zest in what was right for the new one that came so forcefully into being. In time, she found herself a worthy husband, even having children with him, and these she regularly sent to attend the same church wherein they met. Though he never could return there himself, he continued the lovely singing that had first won her heart. She is remembered affectionately, to this day.

Astrological Aspects in Depth (Tenth Lunation)
Major Aspects

Conjunctions: *"Unity, Focus, Intensity"*
These are two planets sitting side by side in the sky, never more than six degrees apart, so they share everything: sign, modality, polarity, and element. Whatever each is on its own becomes intensified, which can be either helpful or challenging, depending on whether they work well or poorly together. A team that is harmonious can accomplish much, but one that always fights guarantees growth and learning (the worse it feels, the greater the motivation). Represented here are two energies united as one, and the fewer degrees that separate them, the harder it can be for someone born with them in her sky to tell the difference between the two. For example, under the twined influence of Mars (great passion) and Uranus (unpredictable changeability within the self) she might think it quite natural and usual for the two to be experienced together and expect that's just how it is, for everyone (which creates a blind spot in her thinking).

271

Sextiles: *"Relaxed, Supportive, Simple, Comfortable"*
What does a sixty degree angle look like? That's how these
planets look in the sky, give or take a maximum of six degrees. Two
signs apart, same polarity, different element: this is a harmonious
arrangement where ideas and opportunities flow easily. Act on
them, and they help you realize your goals. These two planets work
well together, and though they remain apart, they represent things
you can draw upon as needed. These are talents we carry within
ourselves which we may more easily notice, appreciate, and develop.
They won't automatically flow to where they'll do the most good,
you'll need to deliberately direct and guide them, but when they
work together it feels great. Such ease and lack of friction attends
this aspect that it becomes quite tempting to float through life, not
evolve much, and just play, so beware. Dream big, take on
something really huge: you'll have all the support you need.

Squares: *"Struggle, Confusion, Growth, Strength"*
Ninety degrees apart (give or take six at most), and separated by a
sign, this configuration is disharmonious and restless. It is urgent
and even brave, for what we have here are two different elements
at cross purposes from one another that pressure you to work on
the issues they represent, and many changes will be required,
demanding a tremendous effort from you. The many restrictions,
delays, obstacles, and conflicts can be frustrating and very annoying,
but when patience and hard work lead to growth and integration,
the reward is quite sweet. So life is harder when we are younger,
and gets sweeter with every success we win. Squares can only form
between signs whose elements are thought to be incompatible.

Trines: *"Laziness, Natural Talent, Power"*
One-hundred-twenty degrees apart in the sky (give or take no
more than six) and separated by two signs, we have here a passive
configuration that is highly harmonious. Considered most
fortunate of all, this seamless flow of luck, innate talent, and
opportunity – so unstinting it can be taken for granted or even
ignored – brings us talents so natural they are almost unconscious.

272

So, they're not much appreciated, and perhaps never even expressed. It's all about accepting and being accepted, bringing ease to the point of laziness. Here the energy flows naturally into constructive channels, without thought. Why were you born? Is there something really hard you are poised to do? Nothing less lends the pressure you'll need, if you are to evolve.

Oppositions: "Conflict, Motivation, Challenge"
Two planets on opposite sides of the sky (more or less): a very dynamic configuration, even though it is disharmonious. What it denotes is a state of divided loyalties. Deep tension, uncertainty, and insecurity usually result, showing up in the psyche as a dualism in which you must either find the right compromises between two extremes or keep letting each one bounce back and forth while it competes for center stage in your life. It can be intensely frustrating, going on and on as a core experience in life, but through hard work a balance of these opposite feelings can be discovered, by noticing how the two are actually parts of one whole. The oppositions in your chart (if you have any) will encourage you to look for other people who have the right kind of astrological energy, so you can learn about yourself under the action of your interactions with them. Even though this usually creates tension, they only ever occur between signs that are compatible with one another from an elemental point of view.

The above astrological insights (for major aspects) I learned at https://cafeastrology.com, and the following (for minor ones) are from https://astrolibrary.org/aspects-in-astrology:

Minor Aspects:

The Semi-Sextile
Only thirty degrees apart, this aspect is so mild it's almost unnoticeable. Some people see it as mildly beneficial, but since it connects signs of different elements and qualities, there are

times when it can be difficult, awkward, or uncomfortable (in the mildest way), so it's best to interpret it only when taking in the deeper insights afforded by the chart as a whole. Usually, this minor aspect will corroborate with the other things that the chart is saying.

The Semi-Square
Forty-five degrees apart and separated by one eighth of the sky, this is like the square, but less intense. It does imply friction, generating minor internal conflicts or frustrations that are sometimes annoying.

The Sesquiquadrate
One-hundred-thirty-five degrees apart, this configuration is also known as the sesquisquare, or "a square and a half." While it denotes the same kind of friction as the semi-square (correlating with inner conflict) the friction here tends toward external troubles.

The Quincunx
One-hundred-fifty degrees of separation (five twelfths of the sky) demands continual adjustments. Though not necessarily negative, it still brings strain and stress, sometimes so much so that it affects the health, and can either indicate self-neglect regarding the indicated area of life, or circumstances being forced on the person.

The Quintile
Seventy-two degrees apart and the relationship takes up one fifth of the sky: this mildly beneficial configuration implies versatile artistic or creative talents on the mental plane, alongside an authentic style of individuality.

The Biquintile
One-hundred-forty-four degrees apart: like the quintile, only milder.

More Horn Dance (Twelfth Lunation)

I wondered when and how the **Horn Dance** in Abbots Bromley came to be moved from the winter solstice to early fall, so I went back to the web and learned a lot (though never quite what I was looking for), both from David Parr at California Revels and Andrew Bullen at Country Dance and Song.

The earliest written reference to the Horn Dance was penned in 1686, when, in *A Natural History of Staffordshire* (where the village is located), Dr. Robert Plot described a dance he called "the hobby horse dance," involving six sets of caribou horns and a hobby horse. It was done at Christmas, in 1532. It has been suggested that the custom might date as far back as August of 1226, when the Barthelmy Fair was held in Staffordshire, England. With nothing written describing its origin, we can't really know how old it is or isn't. All we know is that "somewhere in the 19th Century, the dance was moved from the winter solstice to its current date."

There was more about the horns, too. Three of the six sets are painted white and three painted blue (currently brown, but the earlier coat shows through). In the 1970's, one of the white sets was damaged, and while being repaired underwent carbon 14 dating that showed it to be over nine hundred years old.

How these antlers came to this parish is still a mystery. They are from domesticated caribou. At least one historian has identified them with an eleventh century monk named Wulfric, counselor to King Ethelred and founder of a Benedictine Abbey on whose land the village was founded in 1004. He speculates they were brought in by Vikings. In 1010, Wulfric defended Mercia against Vikings. Could they have been snagged as trophies of war and the first Horn Dance there been a victory dance? That's one theory, anyway. They are kept in the parish hall of the St. Nicholas Church at Abbots Bromley as property of the Abbots Bromley parish council, and are never allowed to leave the Parish. A smaller, lighter set of red deer antlers are used for practice; they also make guest appearances elsewhere.

Apparently the village of Abbots Bromley has a second name: look for it as Pagets Bromley. If you look for it as both, you're bound to find it. As the dancers in this village all came from the one family, they have passed the steps and patterns down. One of the melodies that's come to be associated with the dance is a tune an elderly village wheelwright learned from his grandfather, who played it towards the end of the 18th Century: "The Wheelwright Robinson's Tune" (his own surname - one of the two borne by the family who carries on this dance). Sheet music for this as well as for several other tunes associated with it can be found in volume 17 of Country Dance and Song (issued May of 1987) in an article written by Andrew Bullen. There, too, you will find diagrams of each dance step.

I learn from David Parr at www.californiarevels.org that the Antler Dance ritual "has come to be 'owned' by many other communities outside of Abbots Bromley; celebrants have added their own special grace notes to the stepping and the costuming. Some dance it on Mayday morn..." and a version of it is done regularly at renaissance faires local to me. Even in the heat and the dust, even with frequent repetitions (danced several times daily on faire week-ends) it retained a certain mystery, despite my being regularly present as part of yet another entertainment group there. "...and it has become a staple of traditional music camps where it is commonly danced by women and children. It has been successfully reshaped by varied groups of celebrants as they embrace its essence and tailor the performance to make it their own. In the California Revels, they reserve its performance for the darkest point in the Solstice show, where its haunting power is allowed to take hold through the almost hypnotic interweaving of the figures and tune. The ten men who dance it here employ a slightly vaulting step that is noticeably different from the way it is performed even in other Revels cities. Our Fool plays the triangle, a boy dances the archer, and Maid Marian is invariably a robust, bearded man...There is no logical answer to the question of the meaning of the Abbots Bromley dance. One explanation is as good as another – and ultimately as wrong. It is a mystery that fares better in the realm of experience than understanding. Each of us feels the power of this ritual and we respond from a place that is deeper than the conscious mind can fathom. It is primitive magic and the awareness of the moment draws us together in a wild and powerful way..." Indeed, sir. It does indeed.

Christmas and Yule (12ᵗʰLunation)

In Robert Sass's *The Old Saxon Heathen Calendar* (published online) I learn that "Hakon the Good (died 961 AD) moved Heathen Yule in Norway to be at the same time as Christmas, then held on the Winter Solstice..."

"...Chapter 15 of the saga "Hakon the Good:

King Hakon...made a law that the festival of Yule should begin at the same time as Christian people held it, and that every man, under penalty, should brew a meal of malt into ale, and therewith keep the Yule holy as long as it lasted. Formerly the first night of Yule was Hǫkunótt, that is Midwinter Night (celebrated on the full of the ÆfterraGeola Moon), and was held for three nights...He dwelt long in the Throndhjem district, for the strength of the country lay there; and when he thought that, by the support of some powerful people there, he could set up Christianity he sent a message to England for a bishop and other teachers; and when they arrived in Norway, Hakon made it known that he would proclaim Christianity over all the land. The people of More and Raumsdal referred the matter to the people of Throndhjem..."

Those Lunar Meditations

It was Jessica MacBeth who formulated and then led a group of us through those lunar meditations in the mid 1980's. I have shared them as I experienced and as I remember them, for to do otherwise would be less than authentic, yet I gather they have evolved somewhat since that time. Jesa had wanted to send them to me in their updated form, but in order to find the booklet they are housed in, she would have had to pick her house up and shake it; and alas! None of the folks in her immediate circle had the necessary strength to do that. So we are left with these – less than complete perhaps, yet nevertheless appreciated.

A Celestially Timed Game (12ᵗʰ Lunation)

While double-checking on the words to "The Holly and the Ivy," I was surprised to discover that someone had invented a game. Simple but elegant, it only needs a few people who enjoy using their imaginations and some dice. It's only meant to be played at Mid-winter, and it pairs perfectly with the song. The original post informs us that it "is supported by my patrons at..." patreon website/Meguey, "where I am making a series of short seasonal games you can play with anyone. This..." (The Holly and the Ivy: a Mid-Winter's Game for 2-6) "is the first in that series. The next game will be out May 1st." So interesting! As it turns out, the lady is famous (Meguey Baker by name) and has designed a great many games and game books, including games for social change among girls in Ethiopia(!) I knew nothing about her initially, but found her Holly and The Ivy game so intriguing it inspired me to dust off and tie up loose ends on something of my own that had rolled into my mind one day out of the blue, whole in spirit if misty in detail. It is bardic in style in that it invites people to share stories, poems, jokes, or songs, and appears to be part game and part quest. A quest of what, you wonder? I do not know. Some airy thing - perhaps the moon, or its bright edge.

The Moon's Bright Edge: A Bardic Quest

A wind there blew across the broken sea,
as soft and salty tides spilled on the shore,
carrying ancient water songs to me, whose beauty
left a silent yearning at my core.

This game pairs well with the new moon. It can run anywhere from one to three days, depending on how long people want it to last. People who want to play in sympathy with the new moon can use a luni-solar calendar to find the ideal length of the game and the best time to begin. To do this they would go to the dark moon page relating to the lunation that is about to unfold and count the number of days falling between Waning Starday and the day the new moon emerges. However many they are (not counting the day of emergence) shows the proper length of game time for that lunation.

And when should the game begin? When is the moment of emergence? Anywhere between 6:00 a.m. and 6:00 p.m. and the game starts at sunrise on that same day. Otherwise, it would start at sunset, either the night before (if emergence is in the a.m.) or after (if p.m.)

The game is highly suitable to house parties. A potluck style works well. Setting up the table well before is wise, and each guest should bring a die. You don't need every player to be present all the time. If at least three show up by starting time, that's good enough. The rest may trickle in or out as they please – their places will have already been set.

'Twas wonder drew me, seeking everywhere;
in wind, in wave, in cities down below,
for fey was this vibration, light as air,
and soothing as an angel's quiet glow.

Setting up the Table

Deal out cards for all who have RSVP'd. Start with the center (played by the host/ess), using the whole deck (any type, face down), and work your way around the table until the deck is in equal shares amongst all players. Leftovers are gathered up; any numbered 8 or lower go face down inside a box (the Treasure Box) next to the center (or Dungeon), with the lid left open. The remaining (higher numbered) cards together with the jokers go into the Dungeon. For each card thrown in, one numbered lowest (beginning with deuces) is released and joins those in the Treasure Box, so that ultimately, the Dungeon has an equal share of cards, more high than low, including any jokers, which are wild.

Seat Perilous & What to Do When Seated There

All but the host/ess (Guardian) roll a single die to see who goes first. The lowest roll attracts Seat Perilous, which moves around the circle, always to the left.

The Guardian turns the top card in the Dungeon face up, and the Challenger in Seat Perilous turns up the top card in his or her pile too. The higher card adds the lower to its pile. If this means the Guardian's Dungeon releases a card, Seat Perilous moves on, but if the Challenger's own card is imprisoned, it may now be ransomed.

Look there! A fish arises, hoary scaled,
and says from calm-pooled eyes with mirthful streak:
"Poor lassie," (in a voice as rich as ale),
"not sea nor earth nor sky hold what you seek."

Will the Challenger tell a joke or a story? Will s/he play or sing a song? There is room for original material here, but it is not required. Any favorite creation will do. Will s/he ask the group if s/he can do anyone a favor? It can be as simple as bringing someone a glass of water. If hands are raised in reply and from among them s/he picks one but then turns down the request, s/he must forfeit some article from about his or her person and leave it in the denied one's keeping, where it may be ransomed at a later time or not, as wished. Either way, the losing card too defects to the denied one's pile. And in the main game, if the owner of an imprisoned card doesn't care to ransom it, s/he must sacrifice something of his or hers (even if it's only a hair tie or a penny, or a message that won't be read until much, much later) in addition to the losing card, placing it into the Treasure Box beside the Dungeon where all such will be held until after the game ends.

And Furthermore

It is the Guardian against whom all others play, and there s/he must remain, unless a substitute (from outside the game) is found. All others may leave or return at will. As long as at least three are at table, the game continues, and any pause resumes again as soon as a missing third returns. Should there ever be less than four, the Guardian too must either ransom a losing card or pay a forfeit just like the others, until a fourth returns. Anyone leaving with no plan of returning must take the cards from his or her place and throw them face-down into the Dungeon.

Ending the Game

A game beginning at sunrise ends at sunset on the terminating day; one begun at sunset ends at sunrise. When dawn's light or dusk's darkness demands a lighting change, the game ends and the table is cleared except for the Treasure Box, which stays in the middle. The dice return, and all players with something of theirs inside the Box gather 'round. Everyone rolls a single die to see who goes first. The highest roll takes an item of choice from the Box. The player to the left gets the first player's die and rolls it with his or her own. If they come up doubles s/he chooses something from the Box too, passing one of the die to the player on the left. If not, s/he takes nothing, passes both dice, and the next player rolls all three, hoping for doubles. Success means something is taken and only one die is passed; if not, all in hand pass to be rolled all together, and nothing leaves the Box. And so it goes, three times around the table until the last round is complete. At this point, the Treasure Box is closed. Whatever is inside becomes the property of the Guardian, but any player in the circle may request a Ransom Meet, to be held the following morn (or eve, as best befits).

A Treasure Box Ransom Meet

The table stands, with Treasure Box on top, lid closed. All players gather 'round (including the Guardian, who doesn't roll yet) and roll a single die, as all will do throughout the game. The highest goes first, rolling against the Guardian, who now must ransom or forfeit like anybody else. Items forfeited go into the Box, but the lid is quickly closed again. Play passes not to the left but to the right, as it will do throughout the game. Anyone not willing to either ransom or forfeit for a losing card must drop from the game. If the Guardian drops, possession moves to the one for whom s/he did so and this newest Guardian defends the Treasure.

Ending the Treasure Box Ransom Meet

The game ends either when only one player remains or at sunset (sunrise), whichever comes first. Those who never dropped may now reclaim any items in the box that were theirs originally. Suddenly someone in a blindfold comes in out of nowhere, seeks, finds, and then snatches up the Box! ...and is led away – to sit in a room beside a pile of small bags (exactly as many as there are players). Still wearing the blindfold, s/he puts an equal number of abandoned things in each, making sure cards too are equally divided, and ties them closed. Now all are placed in one large basket, where pens and small papers lie. The blindfold comes off and the Treasure is carried into the game room. Each player selects a bag (keeping it closed), a pen, and a slip. On this s/he writes his or her name, and beside it, a number and a suit (for example 3 spades, or 11 diamonds: the Jack of Diamonds).

Now is the moment of truth. Do any of the cards in the chosen bag match both the number and the suit that the player wrote down? Yes? Add the number on that card to the one formed by adding up all of the numbers in his or her birthday. Remember that number. No? Well, when each number on every card in the bag is added together, does the resulting number at least match the original number written there? No? Try reducing both (like how 25 turns into 2+5, or 7), and then compare. If yes to any of these, the players keep the things in the bags and will need to remember these winning numbers; if not, the bags go back to the host/ess.

Those who lose bags must wait for their luck to find them. A lucky number can be the "bag" that keeps it safe until then. To reveal it, s/he will use the number gotten when adding up and reducing all the numbers on all the cards in the bag s/he must lose and Voila! This is the one. Now, everyone compares the numbers they've formed. The closest to either13 or 31 gets to host the next game. All remaining cards return to the deck.

"Listen though – the place where music lies
(within and past the shadowed face of fear)
goes with you, in your voice and in your eyes."
He left me then, to puzzle what I dared.

Examples of Diana Worship through Time

Diana was known for blessing moms with a safe and easy labor. Her lunar cycles seemed to parallel the menstrual cycle and helped to track the months during pregnancy. Her temple at Aricia also offered care of infants, pups and pregnant dogs, which extended to the training of young people and dogs, especially for hunting. A festival to Diana was held yearly at Lake Nemi, between August 13th and 15th. Worshipers arrived carrying torches and garlands, and once there, they left pieces of thread tied to fences and tablets inscribed with prayers.

Slaves and the lower classes found asylum in Diana (their patroness)'s temples. A heavily guarded tree in the center of her grove at Nemi was reserved for runaway slaves. Only such a one was allowed to try, if he could, to break one of the boughs. Success granted the privilege to engage her current priest-king in one-on-one mortal combat. The survivor won continuance in that post until a new ex-slave defeated him.

According to Frazer, the '*rex Nemorensis*' (the king at Nemi) was the incarnation of a dying and reviving god, a solar deity who participated in a mystical marriage to a goddess. He died at the harvest and was reincarnated in the spring.

Diana of Nemi was not native to Rome and neither was Juno of Veii, but they made Juno a citizen by a rite called evocatio.

Diana remained a foreigner, in a temple of her own outside city bounds, thus keeping her within reach of all, and not exclusively Roman. Diana Aventina however did have her own temple, in the heart of Rome.

A 7th Century List of Forbidden Things

The list includes something relating to "the first work of the day," giving significance to the various phases of the moon, making little deer or vetulas (corn dollies, figures of the Old Woman), setting tables at night (for the house elf), exchanging New Year's gifts, supplying superfluous drinks (especially at Yule), performing solestitia (dancing or leaping around or over fires or chanting songs, especially on summer solstice eve), giving significance to places where three roads meet or making temples or sanctuaries of rocks, springs, groves, or corners, hanging "phylacteries" (charms or amulets) from the neck of man or beast, making "lustrations" (spiritual cleansings, by washing or sprinkling with water, rubbing with blood or clay, confession, scapegoat or fumigation), incantations with herbs, passing cattle through a hollow tree or ditch, hanging amber from one's neck or calling on Minerva in one's weaving or dyeing, and swearing by or calling the sun or moon "Lord". (From *the Vita Eligii,* by Audoin)

In the West

Over 100 inscriptions to Diana have been catalogued in the provinces, mainly from Gaul, Upper Germania, and Britannia. She was commonly invoked alongside another forest god known as Silvanus in what is now Wiesbaden. By the Mattiaci tribe she was worshiped as *Diana Mattiaca.* She was particularly important in the region in and around the Black Forest, where she was worshiped as *Diana Abnoba* (<u>Abnoba</u> was an indigenous, local goddess).

In the Middle Ages

Certain sermons that were written down and other religious documents still extant show evidence of Diana worship in the Middle Ages and on into Merovingian times. It may have been widespread all over Europe, especially among the more remote communities.

In medieval Belgium a natural spring called the "Fons Remacli" may have seen late-surviving Diana worship. Remacle was the head of a monastery at Solignac. He believed he had met with worship given to "Diana of the Ardennes" (Diana and Celtic Arduinna, blended), near the river Warche. It included effigies and "stones of Diana." He thought there was demonic energy in the spring and that this is what made it run dry, so he did an exorcism at the water source. He then installed a lead pipe, which allowed the water to flow again.

Church records of Northern Italy, Western Germany, and Southern France during the Middle Ages all speak of night-time spirit processions led by a female figure. Spirits were said to enter houses and eat food, which miraculously re-appeared. They would sing and dance, giving advice regarding healing herbs and the location of lost objects. If the house was in good order, they brought fertility and plenty. If not, they cursed the family. Historian Carlo Ginzburg has referred to these legendary spirit gatherings as "The Society of Diana."

Some women reported participating in these processions while their bodies still lay in bed. Local clergy complained that women believed they were following Diana or Herodias, riding out on appointed nights to join the processions or carry out instructions from the goddess. By 1310, the names of the goddess figures attached to the legend were sometimes combined as Herodiana. It is likely that the clergy of this time identified the procession's leader as Diana or Herodias so as to fit an older folk belief into a Biblical framework. Herodias was often blended with her daughter Salome in legend, which also holds that upon being presented with the severed head of John the Baptist, she was blown into the air by wind from the saint's mouth, and that she continued to wander through this wind for all eternity.

It is likely that the spirit identified by the Church as Diana or Herodias was called by pre-Christian names like Holda, or Latin Abundia (meaning plenty), Satia (meaning full), and Italian Richella (rich). Whatever her true origin, by the 13th Century, the leader of the legendary spirit procession had come to be firmly identified with Diana and Herodias through the influence of the Church.

Mars in the West

The importance of Mars in establishing religious and cultural identity within the Roman Empire is indicated by the vast number of inscriptions identifying him with a local deity, especially in the West. Many inscriptions were found there to "Mars_____" (insert name of this or that Celtic god here). These names included Camulus, Belatucadros, Teutates, Visucius, Mullo (a name which may mean "Mule" or "Hill" or "Heap"), Nodens (which could mean "Catcher," so a fisher or a hunter), Rigisamus ("*rig*" showed up at the end of certain Gaulish names as "___rix" and surfaced again in later Celtic languages as *rí,* meaning "ruler," so "Supreme Ruler," "King of Kings"), Rigonemetis ("King of the Sacred Grove"), Segomo (the Victorious), and others; such as Condatis (probably related to a Gaulish place name for settlements at the confluence of rivers, he may have functions pertaining to water and healing), Corotiacus (with a depiction as cavalryman, armed and riding a horse which tramples a prostrate enemy beneath its hooves), Olloudius (depicted without armor, as a *Genius* carrying a double cornucopia and holding a libation bowl), Vorocius (a healer of eye afflictions with a shrine at a curative spring, where burnt offerings could be made), Cocidus (beside an identification with Silvanus and including a depiction of Mars as huntsman accompanied by dog and stag), Thincsus (depicted as a god flanked by two goddesses – he *has* been linked with two goddesses called the Alaisiagae – and accompanied by a goose, geese being frequent companions of Celtic war gods), and Neton or Neto (a Celto-Iberian god at Acci or modern Guadix who may be connected to Irish Neit; he wore a radiant crown like a sun god, because the passion to act with valor he experienced as a kind of heat).

Also on this list is Mars Lenus (he had a major healing cult – among the votives are images of children offering doves; his consort Ancamna is also found with the Celtic Smertrios), and last, there was Loucetios (from louk(k)et, which means bright, shining, or flashing, hence also lightning), who was either a common Celtic metaphor between battles and thunderstorms or else the aura of a divinized hero, but either way this epithet too was linked with Mars. His consort Nemetona ("sacred privilege," nemeton, the sacred grove) was identified with the goddess Victoria. I learned all these things about both Diana and Mars through Wikipedia, while exploring the origins of planetary names.

This Wild Magic

It's wild, this magic – through the earth it goes.
It won't be labeled, analyzed or bound.
From hidden places, out like gold it flows.
Between the stars, in lightening, clouds, rainbows,
it dances, joyous face with laughter crowned.
It's wild, this magic – through the earth it goes.

In damp tree roots and buried seeds it sows
it's secret song of being, where peace resounds.
From hidden places out like gold it flows.
Sweet spiral dance, in every cell it glows
like Grail wine, a dream whose grace astounds.
It's wild, this magic – through the earth it goes!

An Index of Contents

Mothlike,
my fantasy arose,
and sailed the moon
through starry seas.
From dusk she sailed
to the dawninglands, pouring
silver through the galaxies.
The Pleiades
wove their nightly dance
through nets of misty radiance,
while star-shimmer
slanting fell:
bright jewels
set in silver heaven's well.
Mighty were the winds that filled her sails!
At last,
astride a comet's tail,
my fantasy returned,
drunk with mystery.

www.ingramcontent.com/pod-product-compliance
Lightning Source LLC
Chambersburg PA
CBHW071138130626
46553CB00004B/1422